Biodiesel

Ayhan Demirbas

Biodiesel

A Realistic Fuel Alternative for Diesel Engines

Ayhan Demirbas
Professor of Energy Technology
Sila Science and Energy
Trabzon
Turkey

ISBN 978-1-84996-696-2 e-ISBN 978-1-84628-995-8

DOI 10.1007/978-1-84628-995-8

British Library Cataloguing in Publication Data
Demirbas, Ayhan
Biodiesel : a realistic fuel alternative for diesel engines
1. Biodiesel fuels
I. Title
662.8'8

Cover design: eStudio Calamar S.L., Girona, Spain

Printed on acid-free paper

9 8 7 6 5 4 3 2 1

springer.com

Preface

This book aims to provide a comprehensive treatment of triglycerides (oils and fats), which convert primary forms of energy into a more usable and economical source of energy for transportation. Biodiesel is a domestic fuel for diesel engines derived from natural oils like soybean oil. It is the name given to a variety of ester-based oxygenated fuels from renewable biological sources that can be made from processed organic oils and fats.

The text is geared toward postgraduates in energy-related studies, fuel engineers, scientists, energy researchers, industrialists, policymakers, and agricultural engineers and assumes the reader has some understanding of the basic concepts of transportation fuels.

The first chapter, "Introduction to energy sources", comprises one fifth of the book; the chapter goes into detail on global energy sources, especially renewables, *i.e.*, biomass, hydro, wind, solar, geothermal, and marine. The second chapter is entitled "Biofuels" and covers the main liquid biofuels such as bioethanol, biomethanol, and liquid fuels from Fischer–Tropsch synthesis. The third chapter, "Vegetable oils and animal fats", covers the use of vegetable oils and animal fats in fuel engines. Furthermore, processing conditions as well as alternative applications of fatty acid methyl esters are discussed briefly in subsequent chapters-"Biodiesel", "Biodiesel from triglycerides via transesterification", "Fuel properties of biodiesels", "Current technologies in biodiesel production", "Engine performance tests", "Global renewable energy and biofuel scenarios", and "The biodiesel economy and biodiesel policy".

Experts suggest that current oil and gas reserves will last only a few more decades. To meet rising energy demands and compensate for diminishing petroleum reserves, fuels such as biodiesel and bioethanol are in the forefront of alternative technologies. It is well known that transport is almost totally dependent on fossil-, particularly petroleum-, based fuels such as gasoline, diesel fuel, liquefied petroleum gas, and natural gas. An alternative fuel to petrodiesel must be technically feasible, economically competitive, environmentally acceptable, and easily available. Accordingly, the viable alternative fuel for compression-ignition engines is biodiesel. Biodiesel use may improve emission levels of some pollutants and worsen that of others. The use of biodiesel will allow for a balance between agriculture, economic development, and the environment.

The manuscript for this text was reviewed by Anthony Doyle and Simon Rees. I would like to thank the publisher's editorial staff, all of whom have been most helpful.

Trabzon, TURKEY, May 2007 *Ayhan Demirbas*

Contents

Chapter 1
Introduction

1.1 Introduction to Energy Sources

Energy is defined as the ability to do work. Energy is found in different forms, such as heat, light, motion, and sound. There are many forms of energy, but they can all be put into two categories: kinetic and potential. Electrical, radiant, thermal, motion, and sound energies are kinetic; chemical, stored mechanical, nuclear, and gravitational energies are potential forms of energy. There are many different ways in which the abundance of energy around us can be stored, converted, and amplified for our use. Energy cannot be seen, only the effects of it are experienced, and so it usually a difficult subject to grasp. For example, thermal energy transfer by radiation and conduction occur by different processes, but the essential differences (*e.g.*, with respect to process speeds) are only rarely appreciated. Similarly, light and electric energies transfer by wave, but they are different processes.

Energy sources can be classified into three groups fossil, renewable, and fissile. The term fossil refers to an earlier geological age. Fossil fuels were formed many years ago and are not renewable. The fossil energy sources are petroleum, coal, bitumens, natural gas, oil shales, and tar sands. The main fissile energy sources are uranium and thorium. Table 1.1 shows the energy reserves of the world (Demirbas, 2006a), and Fig. 1.1 shows worldwide fossil, nuclear, and renewable energy consumption in 2005. Worldwide, petroleum is the largest single source of energy, surpassing coal, natural gas, nuclear, hydro, and renewables (EIA, 2006).

The term fissile applies to materials that are fissionable by neutrons with zero kinetic energy. In nuclear engineering, a fissile material is one that is capable of

Table 1.1 World energy reserves

Deuterium	Uranium	Coal	Shale oil	Crude oil	Natural gas	Tar sands
7.5×10^9	1.2×10^5	320.0	79.0	37.0	19.6	6.1

Each unit = 1×10^{15} MJ = 1.67×10^{11} bbl crude oil.
Source: Demirbas 2006a

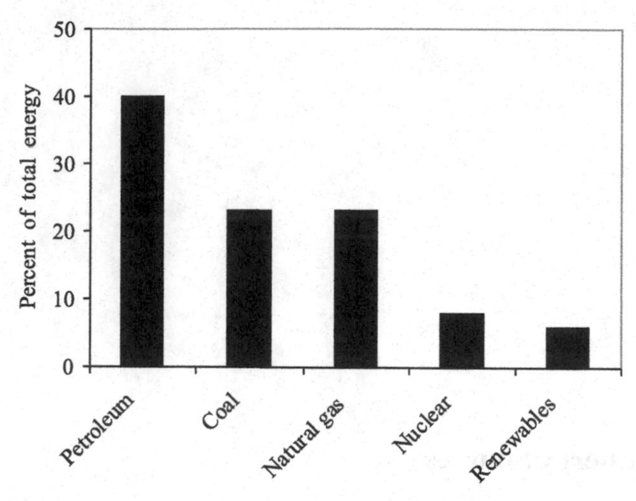

Fig. 1.1 Worldwide fossil, nuclear, and renewable energy consumption (2005)

sustaining a chain reaction of nuclear fission. Nuclear power reactors are mainly fueled with uranium, the heaviest element that occurs in nature in more than trace quantities. The principal fissile materials are uranium-235, plutonium-239, and uranium-233.

Today, most of the energy we use comes from fossil fuels: petroleum, coal, and natural gas. While fossil fuels are still being created today by underground heat and pressure, they are being consumed more rapidly than they are being created. For that reason, fossil fuels are considered non-renewable; that is, they are not replaced as soon as we use them.

The renewable energy sources such as biomass, hydro, wind, solar (thermal and photovoltaic), geothermal, marine, and hydrogen will play an important role in the future. By 2040 approximately half of the global energy supply will come from renewables, and electricity generation from renewables will be more than 80% of the total global electricity supply (EWEA, 2005; EREC, 2006).

Solar and geothermal energy can be used directly for heating. Other energy sources are not directly usable; hence some kind of conversion process must be used to convert the energy into a different form, that is, one of direct utility (Sorensen, 1983). Fossil and renewable energy can be converted into secondary energy sources like electricity and hydrogen.

Renewable resources are more evenly distributed than fossil and nuclear resources, and energy flows from renewable resources are more than three orders of magnitude higher than current global energy use. Today's energy system is unsustainable because of equity issues as well as environmental, economic, and geopolitical concerns that have implications far into the future (UNDP, 2000).

According to the International Energy Agency (IEA), scenarios developed for the USA and the EU indicate that near-term targets of up to 6% displacement of petroleum fuels with biofuels appear feasible using conventional biofuels, given available cropland. A 5% displacement of gasoline in the EU requires about 5% of available cropland to produce ethanol, while in the USA 8% is required. A 5% displacement of diesel requires 13% of US cropland, 15% in the EU (IEA, 2004).

1.2 Global Energy Sources and the Present Energy Situation

Fossil fuels still represent over 80% of total energy supplies in the world today, but the trend toward new energy sources in the future is clear thanks to recent technological developments.

Oil is the fossil fuel that is most in danger of running out. The Middle East is the dominant oil region of the world, accounting for 63% of global reserves. Figure 1.2 shows global oil production scenarios based on today's production. A peak in global oil production may occur between 2015 and 2030. Countries in the Middle East and the Russian Federation hold 70% of the world's dwindling reserves of oil and gas.

The term petroleum comes from the Latin roots *petra*, "rock", and *oleum*, "oil". It is used to describe a broad range of hydrocarbons that are found as gases, liquids, or solids, occurring in nature. The physical properties of petroleum vary greatly. The color ranges from pale yellow through red and brown to black or greenish. The two most common forms are crude oil and natural gas.

Crude oil is a complex mixture that is between 50 and 95% hydrocarbon by weight. The first step in refining crude oil involves separating the oil into different hydrocarbon fractions by distillation. The main fractions of crude oil are given in Table 1.2. Since there are a number of factors that influence the boiling point of a hydrocarbon, these petroleum fractions are complex mixtures.

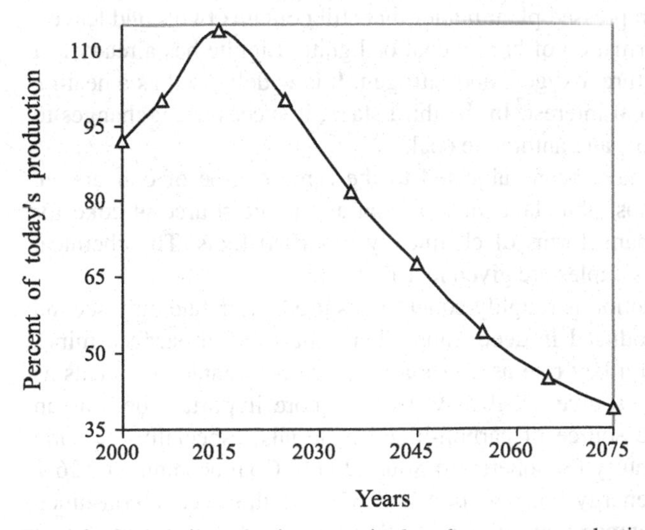

Fig. 1.2 Global oil production scenarios based on current production

Table 1.2 Main crude oil fractions

Fraction	Boiling range (K)	Number of carbon atoms
Natural gas	<295	C_1 to C_4
Petroleum ether	295 to 335	C_5 to C_6
Gasoline	315 to 475	C_5 to C_{12}, but mostly C_6 to C_8
Kerosene	425 to 535	Mostly C_{12} to C_{13}
Diesel fuel	475 to 625	Mostly C_{10} to C_{15}
Fuel oils	>535	C_{14} and higher
Lubricants	>675	C_{20} and above
Asphalt or coke	Residue	Polycyclic

Natural gas (NG) consists mainly of lightweight alkanes, with varying quantities of carbon dioxide, carbon monoxide, hydrogen, nitrogen, and oxygen, and in some cases hydrogen sulfide and possibly ammonia as well. A typical sample of NG, when collected at its source, contains 80% methane (CH_4), 7% ethane (C_2H_6), 6% propane (C_3H_8), 4% butane and isobutane (C_4H_{10}), and 3% pentanes (C_5H_{12}).

The role of NG in the world's energy supply is growing rapidly. NG is the fastest growing primary energy source in the world. The reserves and resources of conventional NG are comparable in size to those of conventional oil, but global gas consumption is still considerably lower than that of oil. Proven gas reserves are not evenly distributed around the globe: 41% of them are in the Middle East and 27% in Russia. A peak in conventional gas production may occur between 2020 and 2050. NG accounts today for 25% of world primary energy production (Jean-Baptiste and Ducroux, 2003). World NG reserves by country are given in Table 1.3 (Yazici and Demirbas, 2001; Demirbas, 2006a).

Coal is basically carbon left over from bacterial action upon decaying plant matter in the absence of oxygen, usually under silt and water. The first step in coal formation yields peat, compressed plant matter that still contains twigs and leaves. The second step is the formation of brown coal or lignite. Lignite has already lost most of the original moisture, oxygen, and nitrogen. It is widely used as a heating fuel but is of little chemical interest. In the third stage, it successively changes to subbituminous, bituminous, and anthracite coal.

Not all coal deposits have been subjected to the same degree of conversion. Bituminous coal is the most abundant form of coal and is the source of coke for smelting, coal tar, and many forms of chemically modified fuels. The chemical properties of typical coal samples are given in Table 1.4.

Worldwide coal production is roughly equal to gas production and only second to that of oil. Coal is produced in deep mines (hard coal) and in surface mines (lignite). Coal has played a key role as a primary source of organic chemicals as well as a primary energy source. Coal may become more important both as an energy source and as the source of carbon-based materials, especially aromatic chemicals, in the 21st century (Schobert and Song, 2002). Coal accounts for 26% of the world's primary energy consumption and 37% of the energy consumed worldwide for electricity generation (Demirbas, 2006a).

Table 1.3 World natural gas reserves by country

Country	Reserves (cubic feet)	Country	Reserves (cubic feet)
Russia	47,573	Pakistan	71
Iran	23,002	India	65
Qatar	14,400	Yugoslavia	48
Saudi Arabia	6,216	Yemen	48
United Arab Emirates	6,000	Brunei	39
United States	5,196	Hungary	37
Algeria	4,500	Thailand	36
Venezuela	4,180	Papua New Guinea	35
Nigeria	3,500	Croatia	34
Iraq	3,100	Bangladesh	30
Turkmenistan	2,860	Burma	28
Australia	2,548	Austria	25
Uzbekistan	1,875	Syria	24
Kazakhstan	1,841	Ireland	20
Netherlands	1,770	Vietnam	19
Canada	1,691	Slovakia	14
Kuwait	1,690	Mozambique	13
Norway	1,246	France	11
Ukraine	1,121	Cameroon	11
Mexico	835	Philippines	10
Oman	821	Afghanistan	10
Argentina	777	Turkey	9
United Kingdom	736	Congo	9
Bolivia	680	Sudan	9
Trinidad and Tobago	665	Tunisia	8
Germany	343	Taiwan	8
Indonesia	262	Namibia	6
Peru	246	Rwanda	6
Italy	229	New Zealand	6
Brazil	221	Bulgaria	6
Malaysia	212	Israel	4
Poland	144	Angola	4
China	137	Equatorial Guinea	4
Libya	131	Japan	4
Azerbaijan	125	Ivory Coat	3
Colombia	122	Ethiopia	3
Ecuador	105	Gabon	3
Romania	102	Ghana	3
Egypt	100	Czech Republic	3
Chile	99	Guatemala	3
Bahrain	91	Albania	3
Denmark	76	Tanzania	2
Cuba	71		

Source: Demirbas, 2006a

Worldwide coal production and consumption in 1998 were 5,043 and 5,014 million short tons, respectively. The known world recoverable coal reserves in 1999 were 1,087 billion short tons (AER, 1999; IEA, 2000). Coal reserves are rather evenly spread around the globe: 25% are in the USA, 16% in Russia, and

Table 1.4 Chemical properties of typical coal samples

	Low-rank coal	High-volatility coal	High-rank coal
Carbon, %	75.2	82.5	90.5
Hydrogen, %	6.0	5.5	4.5
Oxygen, %	17.0	9.6	2.6
Nitrogen, %	1.2	1.7	1.9
Sulfur, %	0.6	0.7	0.5
Moisture, %	10.8	7.8	6.5
Calorific value, MJ/kg	31.4	35.0	36.0

11.5% in China. Although coal is much more abundant than oil and gas on a global scale, coalfields can be depleted on a regional scale.

Nuclear power plants are based on uranium mined in surface mines or by in situ leaching. Nuclear energy has been used to produce electricity for more than half a century. Worldwide, nuclear energy accounts for 6% of energy and 16% of electricity and 23% of electricity in OECD countries (UNDP, 2000). OECD countries produce almost 55% of the world's uranium. Worldwide nuclear energy consumption increased rapidly from 0.1% in 1970 to 7.4% in 1998. This increase was especially high in the 1980s (Demirbas, 2005a).

Renewable energy sources (RESs) contributed 2% of the world's energy consumption in 1998, including 7 exajoules from modern biomass and 2 exajoules for all other renewables (UNDP, 2000). RESs are readily available in nature. Increasing atmospheric concentrations of greenhouse gases increase the amount of heat trapped (or decrease the heat radiated from the Earth's surface), thereby raising the surface temperature of the Earth. RESs are primary energy sources. Renewable energy is a clean or inexhaustible energy like hydrogen energy and nuclear energy. The most important benefit of renewable energy systems is the decrease of environmental pollution.

1.3 Renewable Energy Sources

Renewable energy sources (RESs) are also often called alternative energy sources. RESs that use indigenous resources have the potential to provide energy services with zero or almost zero emissions of both air pollutants and greenhouse gases. Renewable energy technologies produce marketable energy by converting natural materials into useful forms of energy. These technologies use the sun's energy and its direct and indirect effects on the Earth (solar radiation, wind, falling water, and various plants, *i.e.*, biomass), gravitational forces (tides), and the heat of the Earth's core (geothermal) as the resources from which energy is produced (Kalogirou, 2004). Currently, RESs supply 14% of the total world energy demand. Large-scale hydropower supplies 20% of global electricity. Renewable sources are more evenly distributed than fossil and nuclear resources (Demirbas, 2006a). RESs are readily available in nature, and they are primary energy resources.

RESs are derived from those natural, mechanical, thermal, and growth processes that repeat themselves within our lifetime and may be relied upon to produce predictable quantities of energy when required. Renewable technologies like hydro and wind power probably would not have provided the same fast increase in industrial productivity as did fossil fuels (Edinger and Kaul, 2000). The share of RESs is expected to increase very significantly (to 30 to 80% in 2100). Biomass, wind, and geothermal energy are commercially competitive and are making relatively fast progress (Fridleifsson, 2001). In 2005 the distribution of renewable energy consumption as a percentage of total renewable energy in the world was as follows: biomass 46%, hydroelectric 45%, geothermal 6%, wind 2%, and solar 1% (EIA, 2006).

Renewable energy scenarios depend on environmental protection, which is an essential characteristic of sustainable development. World biomass production is estimated at 146 billion metric tons a year, comprised mostly of wild plant growth (Cuff and Young, 1980). Worldwide biomass ranks fourth as an energy source, providing approximately 14% of the world's energy needs (Hall *et al.*, 1992). Biomass now represents only 3% of primary energy consumption in industrialized countries. However, much of the rural population in developing countries, which represents about 50% of the world's population, relies on biomass, mainly in the form of wood, for fuel (Ramage and Scurlock, 1996).

About 98% of carbon emissions result from fossil fuel combustion. Reducing the use of fossil fuels would considerably reduce the amount of carbon dioxide produced, as well as reducing the levels of the pollutants. Indeed, much of the variation in cost estimates to control carbon emissions revolves around the availability and cost of carbon-free technologies and carbon-reducing technologies, such as energy efficiency and energy conservation equipment. This can be achieved by either using less energy altogether or using alternative energy resources. Much of the current effort to control such emissions focuses on advancing technologies that emit less carbon (*e.g.*, high-efficiency combustion) or no carbon such as nuclear, hydrogen, solar, wind, geothermal, or other RESs or on using energy more efficiently and on developing innovative technologies and strategies to capture and dispose of carbon dioxide emitted during fossil fuel combustion. The main RESs and their usage forms are given in Table 1.5 (Demirbas, 2005b).

Renewable energy is a promising alternative solution because it is clean and environmentally safe. RESs also produce lower or negligible levels of greenhouse gases and other pollutants as compared with the fossil energy sources they replace. Table 1.6 shows the global renewable energy scenario by 2040. Approximately half of the global energy supply will come from renewables in 2040, according to the European Renewable Energy Council (2006). The most significant developments in renewable energy production will be observed in photovoltaics (from 0.2 to 784 Mtoe) and wind energy (from 4.7 to 688 Mtoe) between 2001 and 2040.

Table 1.5 Main renewable energy sources and their usage forms

Energy source	Energy conversion and usage options
Hydropower	Power generation
Modern biomass	Heat and power generation, pyrolysis, gasification, digestion
Geothermal	Urban heating, power generation, hydrothermal, hot dry rock
Solar	Solar home system, solar dryers, solar cookers
Direct solar	Photovoltaics, thermal power generation, water heaters
Wind	Power generation, wind generators, windmills, water pumps
Wave	Numerous designs
Tidal	Barrage, tidal stream

Table 1.6 Global renewable energy scenario by 2040

	2001	2010	2020	2030	2040
Total consumption (million ton oil equivalent)	10,038	10,549	11,425	12,352	13,310
Biomass	1,080	1,313	1,791	2,483	3,271
Large hydro	22.7	266	309	341	358
Geothermal	43.2	86	186	333	493
Small hydro	9.5	19	49	106	189
Wind	4.7	44	266	542	688
Solar thermal	4.1	15	66	244	480
Photovoltaic	0.2	2	24	221	784
Solar thermal electricity	0.1	0.4	3	16	68
Marine (tidal/wave/ocean)	0.05	0.1	0.4	3	20
Total renewable energy sources	1,365.5	1,745.5	2,694.4	4,289	6,351
Renewable energy source contribution (%)	13.6	16.6	23.6	34.7	47.7

1.3.1 Biomass Energy and Biomass Conversion Technologies

The term biomass (Greek, bio, life + maza or mass) refers to wood, short-rotation woody crops, agricultural wastes, short-rotation herbaceous species, wood wastes, bagasse, industrial residues, waste paper, municipal solid waste, sawdust, biosolids, grass, waste from food processing, aquatic plants and algae animal wastes, and a host of other materials. Biomass is the name given to all the Earth's living matter. Biomass as solar energy stored in chemical form in plant and animal materials is among the most precious and versatile resources on Earth. It is a rather simple term for all organic materials that derive from plants, trees, crops, and algae. The components of biomass include cellulose, hemicelluloses, lignin, extractives, lipids, proteins, simple sugars, starches, water, hydrocarbons, ash, and other compounds. Two larger carbohydrate categories that have significant value are cellulose and hemicelluloses (holocellulose). The lignin fraction consists of non-sugar-type molecules.

Wood and other forms of biomass are one of the main RESs available and provide liquid, solid, and gaseous fuels. Animal wastes are another significant potential biomass source for electricity generation and, like crop residues, have many applications, especially in developing countries. Biomass is simply an organic petroleum substitute that is renewable (Garg and Datta, 1998; Demirbas, 2004a).

Biomass is the name given to the plant matter that is created by photosynthesis in which the sun's energy converts water and CO_2 into organic matter. Thus biomass materials are directly or indirectly a result of plant growth. These include firewood plantations, agricultural residues, forestry residues, animal wastes, *etc.* Fossil fuels can also be termed biomass since they are the fossilized remains of plants that grew millions of years ago. Worldwide biomass ranks fourth as an energy source, providing *ca.* 14% of the world's energy needs, while in many developing countries its contribution ranges from 40 to 50% (McGowan, 1991; Hall *et al.*, 1992). The use of biomass as fuel helps to reduce greenhouse gas emissions because the CO_2 released during the combustion or conversion of biomass into chemicals is the same CO_2 that is removed from the environment by photosynthesis during the production of biomass.

The basic structure of all woody biomass consists of three organic polymers: cellulose, hemicelluloses, and lignin in the trunk, foliage, and bark. Three structural components are cellulose, hemicelluloses and lignin which have rough formulae as $CH_{1.67}O_{0.83}$, $CH_{1.64}O_{0.78}$, and $C_{10}H_{11}O_{3.5}$, respectively (Demirbas, 2000a). Added to these materials are extractives and minerals or ash. The proportion of these wood constituents varies between species, and there are distinct differences between hardwoods and softwoods. Hardwoods have a higher proportion of cellulose, hemicelluloses, and extractives than softwoods, but softwoods have a higher proportion of lignin. In general, hardwoods contain about 43% cellulose, 22% lignin, and 35% hemicelluloses while softwoods contain about 43% cellulose, 29% lignin, and 28% hemicelluloses (on an extractive-free basis) (Rydholm, 1965).

The main components of lignocellosic biomass are cellulose, hemicelluloses, and lignin. Cellulose is a remarkable pure organic polymer, consisting solely of units of anhydroglucose held together in a giant straight-chain molecule. Cellulose must be hydrolyzed into glucose before fermentation to ethanol. Conversion efficiencies of cellulose into glucose may be dependent on the extent of chemical and mechanical pretreatments to structurally and chemically alter the pulp and paper mill wastes. The method of pulping, the type of wood, and the use of recycled pulp and paper products also could influence the accessibility of cellulose to cellulase enzymes (Adeeb, 2004). Hemicelluloses (arabinoglycuronoxylan and galactoglucomammans) are related to plant gums in composition and occur in much shorter molecule chains than cellulose. The hemicelluloses, which are present in deciduous woods chiefly as pentosans and in coniferous woods almost entirely as hexosanes, undergo thermal decomposition very readily. Hemicelluloses are derived mainly from chains of pentose sugars and act as the cement material holding together the cellulose micelles and fiber (Theander, 1985). Lignins are polymers of aromatic compounds. Their function is to provide structural strength, seal water-conducting systems that link the roots to the leaves, and protect plants against degradation (Glasser, 1985). Lignin is a macromolecule that consists of alkylphenols and has a complex three-dimensional structure. Lignin is covalently linked with xylans in the case of hardwoods and with galactoglucomannans in softwoods. Though mechanically cleavable to a relatively low molecular weight, lignin is not soluble in water. It is generally accepted that free phenoxyl radicals are formed by thermal decomposition of lignin above 525 K and that the radicals have a random tendency to form a solid residue through condensation or repolymerization (Demirbas, 2000b). Cellulose is insoluble in most solvents and has a low accessibility to acid and enzymatic hydrolysis. Hemicelluloses are largely soluble in alkalis and as such are more easily hydrolyzed.

Solar energy, which is stored in plants and animals, or in the wastes that they produce, is called biomass energy. Biomass energy is a variety of chemical energy. This energy can be recovered by burning biomass as a fuel. Direct combustion is the old way of using biomass. Biomass thermochemical conversion technologies such as pyrolysis, liquefaction, and gasification are certainly not the most important options at present; combustion is responsible for over 97% of the world's bioenergy production (Demirbas, 2004a). The average of biomass energy is produced from wood and wood wastes (64%), followed by solid waste (24%), agricultural waste (5%), and landfill gases (5%) (Demirbas, 2000c). Biomass can be economically produced with minimal or even positive environmental impacts through perennial crops.

Biomass has been recognized as a major world RES to supplement declining fossil fuel resources (Ozcimen and Karaosmanoglu, 2004; Jefferson, 2006). Biomass is the most important RES in the world. Biomass power plants have advantages over fossil fuel plants because their pollution emissions are lower. Energy from biomass fuels is used in the electric-utility, lumber and wood products, and pulp and paper industries. Wood fuel is a RES, and its importance will increase in the future. Biomass can be used directly or indirectly by converting it into a liquid or gaseous fuel. A large number of research projects in the field of thermochemical conversion of biomass, mainly on liquefaction, pyrolysis and on gasification, have been performed (Demirbas, 2000a).

When biomass is used directly in an energy application without chemical processing, it is combusted. Conversion may be effected by thermochemical, biological, or chemical processes. These may be categorized as follows: direct combustion, pyrolysis, gasification, liquefaction, supercritical fluid extraction, anaerobic digestion, fermentation, acid hydrolysis, enzyme hydrolysis, and esterification. Figure 1.3 shows the main biomass conversion processes. Biomass can be converted into biofuels such as bioethanol and biodiesel and thermochemical conversion products such as syn-oil, bio-syngas, and biochemicals. Bioethanol is a fuel derived from renewable sources of feedstock, typically plants such as wheat, sugar beet, corn, straw, and wood. Bioethanol is a petrol additive/substitute.

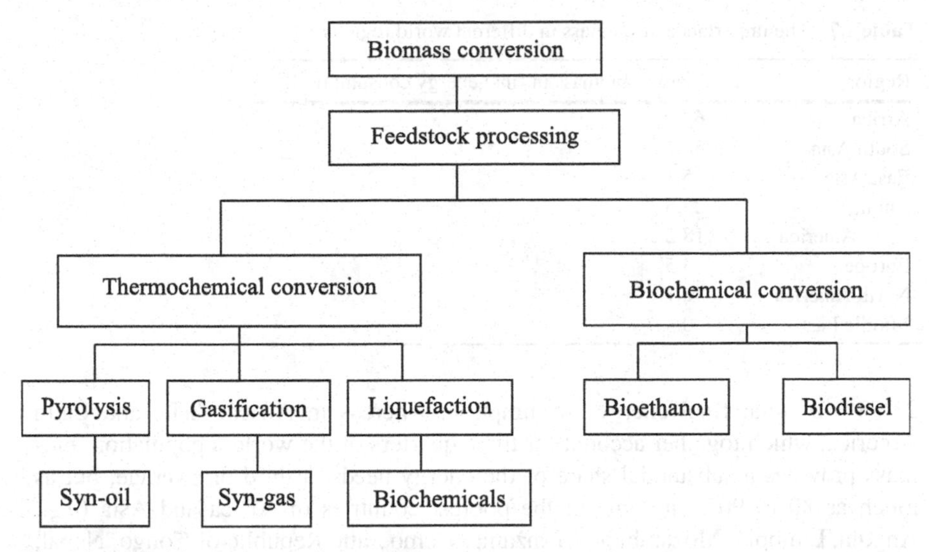

Fig. 1.3 Main biomass conversion processes

Biodiesel is better than diesel fuel in terms of sulfur content, flash point, aromatic content, and biodegradability (Bala, 2005).

Direct combustion and cofiring with coal for electricity production from biomass has been found to be a promising method for application in the nearest future. The supply is dominated by traditional biomass used for cooking and heating, especially in rural areas of developing countries. Traditional biomass from cooking and heating produces high levels of pollutants.

Biomass energy currently represents *ca.* 14% of world final energy consumption, a higher share than that of coal (12%) and comparable to those of gas (15%) and electricity (14%). Biomass is the main source of energy for many developing countries, and most of it is non-commercial. Hence there is an enormous difficulty in collecting reliable biomass energy data. Yet good data are essential for analyzing tendencies and consumption patterns, for modeling future trends, and for designing coherent strategies (Demirbas, 2005b).

The energy dimension of biomass use is importantly related to the possible increased use of this source as a critical option in addressing the global warming issue. Biomass as an energy source is generally considered completely CO_2 neutral. The underlying assumption is that the CO_2 released into the atmosphere is matched by the amount used in its production. This is true only if biomass energy is sustainably consumed, *i.e.*, the stock of biomass does not diminish in time. This may not be the case in many developing countries.

The importance of biomass in different world regions is given in Table 1.7. As shown in this table, the importance of biomass varies significantly across regions. In Europe, North America, and the Middle East, the share of biomass averages

Table 1.7 The importance of biomass in different world regions

Region	Share of biomass in final energy consumption
Africa	60.0
South Asia	56.3
East Asia	25.1
China	23.5
Latin America	18.2
Europe	3.5
North America	2.7
Middle East	0.3

2 to 3% of total final energy consumption, whereas in Africa, Asia, and Latin America, which together account for three quarters of the world's population, biomass provides a substantial share of the energy needs: a third on average, but as much as 80 to 90% in some of the poorest countries of Africa and Asia (*e.g.*, Angola, Ethiopia, Mozambique, Tanzania, Democratic Republic of Congo, Nepal, and Myanmar). Indeed, for large portions of the rural populations of developing countries, and for the poorest sections of urban populations, biomass is often the only available and affordable source of energy for basic needs such as cooking and heating (Demirbas, 2005b).

Biomass is burned by direct combustion to produce steam, the steam turns a turbine, and the turbine drives a generator, producing electricity. Gasifiers are used to convert biomass into a combustible gas (biogas). The biogas is then used to drive a high-efficiency, combined-cycle gas turbine (Dogru *et al.*, 2002). Biomass consumption for electricity generation has been growing sharply in Europe since 1996, with 1.7% of power generation in 1996.

There are three ways to use biomass. It can be burned to produce heat and electricity, changed to gaslike fuels such as methane, hydrogen, and carbon monoxide, or converted into a liquid fuel. Liquid fuels, also called biofuels, include mainly two forms of alcohol: ethanol and methanol. The most commonly used biofuel is ethanol, which is produced from sugarcane, corn, and other grains. A blend of gasoline and ethanol is already used in cities with high levels of air pollution.

1.3.1.1 Liquefaction of Biomass

Recent studies have focused on determining the compounds in oil and aqueous phases obtained from liquefaction processes applied to various raw materials such as biobasic wastes (Qu *et al.*, 2003; Taner *et al.*, 2004). Processes relating to the liquefaction of biomass are based on the early research of Appel *et al.* (1971). These researchers reported that a variety of biomass such as agricultural and civic wastes could be converted, partially, into a heavy oil-like product by reaction with water and carbon monoxide/hydrogen in the presence of sodium carbonate. The heavy oil obtained from the liquefaction process is a viscous tarry lump, which sometimes caused problems in handling. For this purpose, some organic solvents were added to the reaction system. These processes require high temperature and pressure.

In the liquefaction process, biomass is converted into liquefied products through a complex sequence of physical structure and chemical changes. The feedstock of liquefaction is usually wet matter. In liquefaction, biomass is decomposed into small molecules. These small molecules are unstable and reactive and can repolymerize into oily compounds with a wide range of molecular weight distribution (Demirbas, 2000a).

Liquefaction can be accomplished directly or indirectly. Direct liquefaction involves rapid pyrolysis to produce liquid tars and oils and/or condensable organic vapors. Indirect liquefaction involves the use of catalysts to convert non-condensable, gaseous products of pyrolysis or gasification into liquid products. The liquefaction of biomass has been investigated in the presence of solutions of alkalis (Eager *et al.*, 1982; Boocock *et al.*, 1982; Beckman and Boocock, 1983; Demirbas, 1991, 1994), formates of alkaline metals (Hsu and Hisxon, 1981), propanol and butanol (Ogi and Yokoyama, 1993), and glycerine (Demirbas, 1985, 1994, 1998; Kucuk and Demirbas, 1993), or by direct liquefaction (Ogi *et al.*, 1985; Minowa *et al.*, 1994).

Alkali salts, such as sodium carbonate and potassium carbonate, can degrade cellulose and hemicelluloses into smaller fragments. The degradation of biomass into smaller products mainly proceeds by depolymerization and deoxygenation. In the liquefaction process, the amount of solid residue increases in proportion to the lignin content. Lignin is a macromolecule that consists of alkylphenols and has a complex three-dimensional structure. It is generally accepted that free phenoxyl radicals are formed by thermal decomposition of lignin above 525 K and that the radicals have a random tendency to form a solid residue through condensation or repolymerization (Demirbas, 2000b).

The changes that take place during the liquefaction process involve all kinds of processes such as solvolysis, depolymerization, decarboxylation, hydrogenolysis, and hydrogenation. Solvolysis results in micellarlike substructures of the biomass. The depolymerization of biomass leads to smaller molecules and to new molecular rearrangements through dehydration and decarboxylation. When hydrogen is present, the hydrogenolysis and hydrogenation of functional groups such as hydroxyl groups, carboxyl groups, and keto groups also occur (Chornet and Overend, 1985).

Table 1.8 shows the yields of liquefaction products from non-catalytic runs.

Table 1.9 shows the yields of liquefaction products from NaOH catalytic runs.

Figure 1.4 shows the yields of liquefaction products obtained from direct glycerol liquefaction of wood powder. The figure also shows how the yields of liquefaction products increase from 42.0 to 82.0% when the liquefaction temperature is increased from 440 to 560 K. Figure 1.5 shows the yields of liquefaction products obtained from alkali glycerol liquefaction of wood powder in the presence of 5% Na_2CO_3. Figure 1.6 shows the yields of liquefaction products obtained from alkali glycerol liquefaction of wood powder in the presence of 5% NaOH. Figure 1.7 shows the procedures for the separation of liquefaction products.

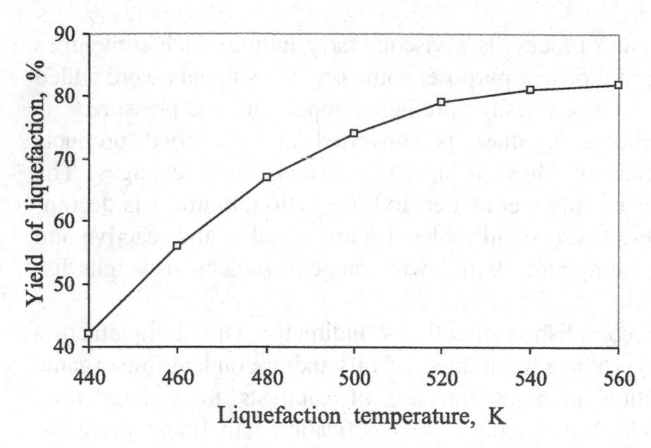

Fig. 1.4 Direct glycerol liquefaction of wood powder

Table 1.8 Yields of liquefaction products from non-catalytic runs (sample-to-solvent ratio: 10 g wood/100 ml water, liquefaction time: 25 min)

Products	550 K	575 K	600 K	625 K	650 K
Water solubles	11.8–13.1	12.6–14.2	14.0–15.4	14.8–16.3	16.7–18.5
Acetone solubles	13.8–16.8	16.3–19.1	19.4–22.8	20.2–22.6	25.8–28.4
Acetone insolubles	28.6–36.2	24.0–32.4	22.6–30.8	19.7–28.7	19.1–28.2

Table 1.9 Yields of liquefaction products from NaOH catalytic runs (catalyst-to-sample ratio: 1/5, sample-to-solvent ratio: 10 g wood/100 ml water, liquefaction time: 25 min)

Products	550 K	575 K	600 K	625 K	650 K
Water solubles	22.8–24.1	26.5–28.6	28.0–30.4	30.8–32.0	32.4–34.5
Acetone solubles	35.7–38.2	39.2–41.5	41.4–43.8	42.8–45.7	46.3–49.4
Acetone insolubles	22.5–24.1	19.2–20.4	17.4–18.6	16.2–17.1	14.5–15.8

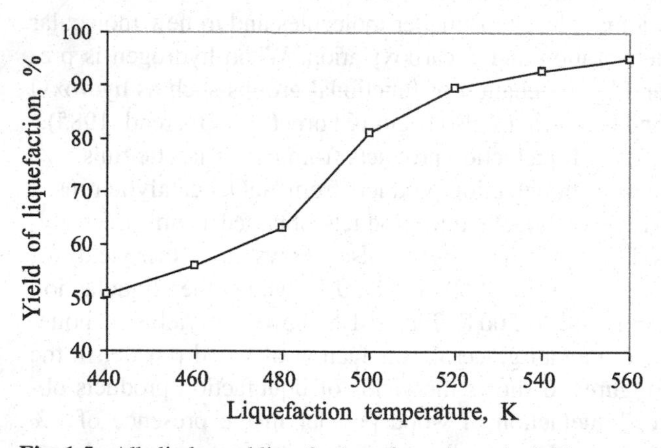

Fig. 1.5 Alkali glycerol liquefaction of wood powder in the presence of 5% Na_2CO_3

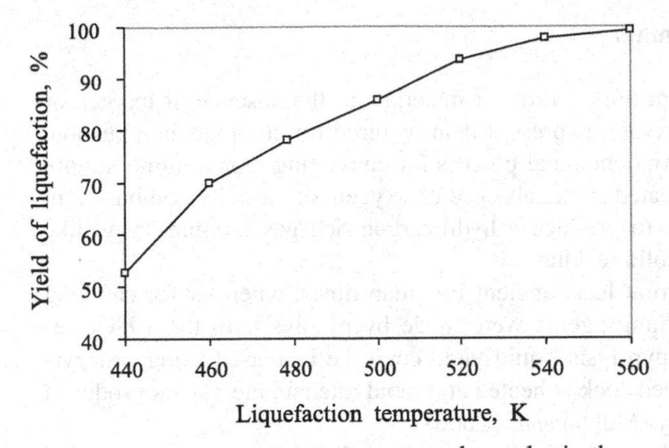

Fig. 1.6 Alkali glycerol liquefaction of wood powder in the presence of 5% NaOH

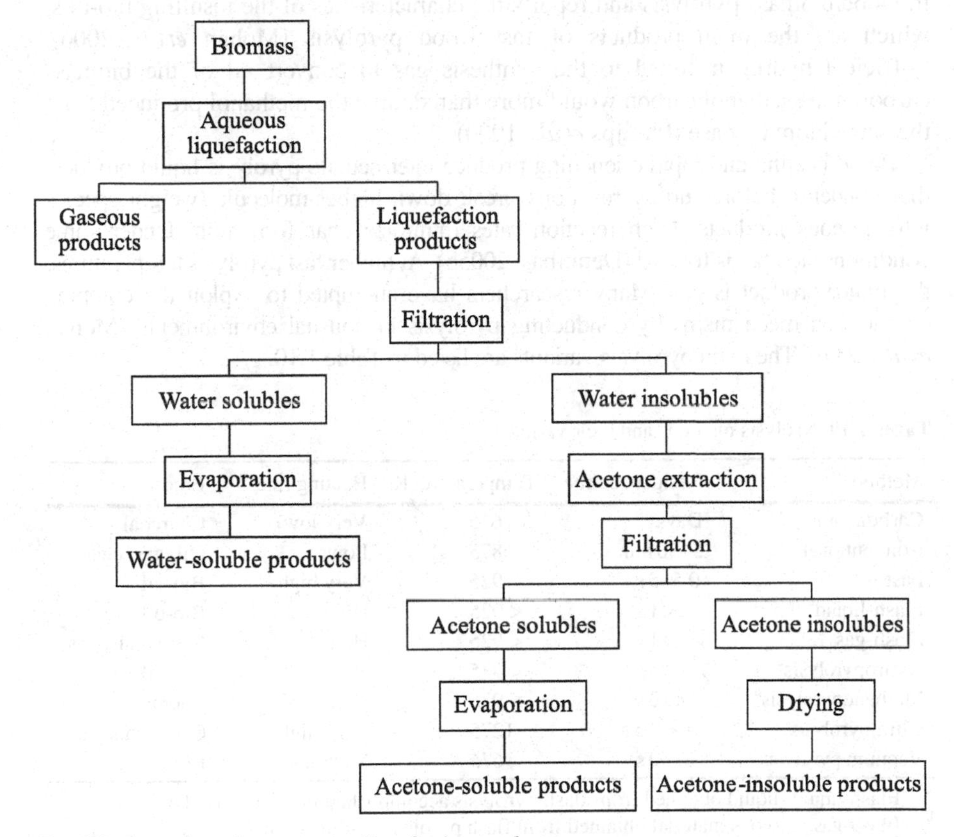

Fig. 1.7 Procedures for separating liquefaction products

1.3.1.2 Pyrolysis of Biomass

Pyrolysis is the thermal decomposition of materials in the absence of oxygen or when significantly less oxygen is present than required for complete combustion. Pyrolysis is the basic thermochemical process for converting biomass into a more useful fuel. Biomass is heated in the absence of oxygen, or partially combusted in a limited oxygen supply, to produce a hydrocarbon-rich gas mixture-an oil-like liquid and a carbon-rich solid residue.

Pyrolysis dates back to at least ancient Egyptian times, when tar for caulking boats and certain embalming agents were made by pyrolysis. In the 1980s, researchers found that the pyrolysis liquid yield could be increased using fast pyrolysis where a biomass feedstock is heated at a rapid rate and the vapors produced are also condensed rapidly (Mohan *et al.*, 2006).

In wood-derived pyrolysis oil, specific oxygenated compounds are present in relatively large amounts (Rowell and Hokanson, 1979; Phillips *et al.*, 1990; Mohan *et al.*, 2006). A current comprehensive review focuses on the recent developments in wood/biomass pyrolysis and reports the characteristics of the resulting bio-oils, which are the main products of fast wood pyrolysis (Mohan *et al.*, 2006). Sufficient hydrogen added to the synthesis gas to convert all of the biomass carbon into methanol carbon would more than double the methanol produced from the same biomass base (Phillips *et al.*, 1990).

Rapid heating and rapid quenching produce intermediate pyrolysis liquid products that condense before further reactions break down higher-molecular-weight species into gaseous products. High reaction rates minimize char formation. Under some conditions, no char is formed (Demirbas, 2005b). At higher fast pyrolysis temperatures, the major product is gas. Many researchers have attempted to exploit the complex degradation mechanisms by conducting pyrolysis in unusual environments (Mohan *et al.*, 2006). The main pyrolysis variants are listed in Table 1.10.

Table 1.10 Pyrolysis methods and their variants

Method	Residence time	Temperature, K	Heating rate	Products
Carbonation	Days	675	Very low	Charcoal
Conventional	5–30 min	875	Low	Oil, gas, char
Fast	0.5–5 s	925	Very high	Bio-oil
Flash-liquid[a]	< 1 s	< 925	High	Bio-oil
Flash-gas[b]	< 1 s	< 925	High	Chemicals, gas
Hydropyrolysis[c]	< 10 s	< 775	High	Bio-oil
Methanopyrolysis[d]	< 10 s	> 975	High	Chemicals
Ultra pyrolysis[e]	< 0.5 s	1275	Very high	Chemicals, gas
Vacuum pyrolysis	2–30 s	675	Medium	Bio-oil

[a] Flash-liquid: liquid obtained from flash pyrolysis accomplished in a time of < 1 s.
[b] Flash-gas: gaseous material obtained from flash pyrolysis within a time of < 1 s.
[c] Hydropyrolysis: pyrolysis with water.
[d] Methanopyrolysis: pyrolysis with methanol.
[e] Ultra pyrolysis: pyrolysis with very high degradation rate.

Pyrolysis is the simplest and almost certainly the oldest method of processing one fuel in order to produce a better one. Pyrolysis can also be carried out in the presence of a small quantity of oxygen (gasification), water (steam gasification), or hydrogen (hydrogenation). One of the most useful products is methane, which is a suitable fuel for electricity generation using high-efficiency gas turbines.

Cellulose and hemicelluloses form mainly volatile products on heating due to the thermal cleavage of the sugar units. The lignin forms mainly char since it is not readily cleaved to lower-molecular-weight fragments. The progressive increase in the pyrolysis temperature of the wood leads to the release of the volatiles, forming a solid residue that is different chemically from the original starting material (Demirbas, 2000a). Cellulose and hemicelluloses initially break into compounds of lower molecular weight. This forms an "activated cellulose" that decomposes by two competing reactions: one forming volatiles (anhydrosugars) and the other char and gases. The thermal degradation of activated cellulose and hemicelluloses to form volatiles and char can be divided into categories depending on the reaction temperature. In a fire all these reactions take place concurrently and consecutively. Gaseous emissions are predominantly a product of pyrolitic cracking of the fuel. If flames are present, fire temperatures are high, and more oxygen is available from thermally induced convection.

Biomass pyrolysis is attractive because solid biomass and wastes can be readily converted into liquid products. These liquids, as crude bio-oil or slurry of charcoal with water or oil, have advantages in transport, storage, combustion, retrofitting, and flexibility in production and marketing.

The pyrolysis of biomass is a thermal treatment that results in the production of charcoal, liquid, and gaseous products. Among the liquid products, methanol is one of the most valuable products. The liquid fraction of pyrolysis products consists of two phases: an aqueous phase containing a wide variety of organo-oxygen compounds of low molecular weight and a non-aqueous phase containing insoluble organics of high molecular weight. This phase is called tar and is the product of greatest interest. In one study, the ratios of acetic acid, methanol, and acetone of the aqueous phase were higher than those of the non-aqueous phase. The point where the cost of producing energy from fossil fuels exceeds the cost of biomass fuels had been reached. With a few exceptions, energy from fossil fuels will cost more money than the same amount of energy supplied through biomass conversion (Demirbas, 2007).

Table 1.11 shows the fuel properties of diesel, biodiesel, and biomass pyrolysis oil. The kinematic viscosity of pyrolysis oil varies from as low as 11 cSt to as high as 115 mm^2/s (measured at 313 K) depending on the nature of the feedstock, the temperature of the pyrolysis process, thermal degradation degree and catalytic cracking, water content of the pyrolysis oil, amount of light ends that have collected, and the pyrolysis process used. Pyrolysis oils typically have water contents of 15 to 30%wt. of the oil mass, which cannot be removed by conventional methods like distillation. Phase separation may partially occur above certain water contents. The water content of pyrolysis oils contributes to their low energy density, lowers the flame temperature of the oils, leads to ignition difficulties, and,

Table 1.11 Fuel properties of diesel, biodiesel, and biomass pyrolysis oil

Property	Test method	ASTM D975 (diesel)	ASTM D6751 (biodiesel, B100)	Pyrolysis oil (bio-oil)
Flash point	D 93	325 K min	403 K	–
Water and sediment	D 2709	0.05 max %vol.	0.05 max %vol.	0.01–0.04
Kinematic viscosity (at 313 K)	D 445	1.3–4.1 mm^2/s	1.9–6.0 mm^2/s	25–1000
Sulfated ash	D 874	–	0.02 max %wt.	–
Ash	D 482	0.01 max %wt.	–	0.05–0.01%wt.
Sulfur	D 5453	0.05 max %wt.	–	–
Sulfur	D 2622/129	–	0.05 max %wt.	0.001–0.02%wt.
Copper strip corrosion	D 130	No 3 max	No 3 max	–
Cetane number	D 613	40 min	47 min	–
Aromaticity	D 1319	–	35 max %vol.	–
Carbon residue	D 4530	–	0.05 max %mass	0.001–0.02%wt.
Carbon residue	D 524	0.35 max %mass	–	–
Distillation temp (90% volume recycle)	D 1160	555 K min –611 K max	–	–

when preheating the oil, can lead to premature evaporation of the oil and resultant injection difficulties. The higher heating value (HHV) of pyrolysis oils is below 26 MJ/kg (*vs.* 42 to 45 MJ/kg for conventional petroleum fuel oils). In contrast to petroleum-based oils, which are non-polar and in which water is insoluble, biomass oils are highly polar and can readily absorb over 35% water (Demirbas, 2007).

Pyrolysis oil (bio-oil) from wood is typically a liquid, almost black through dark red brown. The density of the liquid is about 1200 kg/m^3, which is higher than that of fuel oil and significantly higher than that of the original biomass. Bio-oils typically have water contents of 14 to 33%wt., which cannot be removed by conventional methods like distillation. Phase separation may occur above certain water contents. The HHV is below 27 MJ/kg (*vs.* 43 to 46 MJ/kg for conventional fuel oils).

The bio-oil formed at 725 K contains high concentrations of compounds such as acetic acid, 1-hydroxy-2-butanone, 1-hydroxy-2-propanone, methanol, 2,6-dimethoxyphenol, 4-methyl-2,6-dimetoxyphenol, 2-Cyclopenten-1-one, *etc.* A significant characteristic of bio-oils is the high percentage of alkylated compounds, especially methyl derivatives. As the temperature increases, some of these compounds are transformed via hydrolysis (Kuhlmann *et al.*, 1994). The formation of unsaturated compounds from biomass materials generally involves a variety of reaction pathways such as dehydration, cyclization, Diels-Alder cycloaddition reactions, and ring rearrangement. For example, 2,5-hexandione can undergo cyclization under hydrothermal conditions to produce 3-methyl-2-cyclopenten-1-one with very high selectivity of up to 81% (An *et al.*, 1997).

1.3.1.3 Reaction Mechanism of Pyrolysis

The general changes that occur during pyrolysis are as follows (Babu and Chaurasia 2003; Mohan *et al.*, 2006):

1. Heat transfer from a heat source increases the temperature inside the fuel.
2. The initiation of primary pyrolysis reactions at this higher temperature releases volatiles and forms char.
3. The flow of hot volatiles toward cooler solids results in heat transfer between hot volatiles and cooler unpyrolyzed fuel.
4. Condensation of some of the volatiles in the cooler parts of the fuel, followed by secondary reactions, can produce tar.
5. Autocatalytic secondary pyrolysis reactions proceed while primary pyrolytic reactions simultaneously occur in competition.
6. Further thermal decomposition, reforming, water-gas-shift reactions, radical recombination, and dehydration can also occur, which are a function of the process's residence time/temperature/pressure profile.

A comparison of pyrolysis, ignition, and combustion of coal and biomass particles reveals the following:

1. Pyrolysis starts earlier for biomass compared with coal.
2. The VM content of biomass is higher compared with that of coal.
3. The fractional heat contribution by VM in biomass is on the order of 70 compared with 36% for coal.
4. Biomass char has more O_2 compared with coal. The fractional heat contribution by biomass is on the order of 30% compared with 70% for coal.
5. The heating value of volatiles is lower for biomass compared with that of coal.
6. Pyrolysis of biomass chars mostly releases CO, CO_2, and H_2O.
7. Biomass has ash that is more alkaline in nature, which may aggravate fouling problems.

The organic compounds from biomass pyrolysis are the following groups:

1. A gas fraction containing CO, CO_2, some hydrocarbons, and H_2.
2. A condensable fraction containing H_2O and low-molecular-weight organic compounds (aldehydes, acids, ketones, and alcohols).
3. A tar fraction containing higher-molecular-weight sugar residues, furan derivatives, phenolic compounds, and airborne particles of tar and charred material that form smoke.

The mechanism of pyrolysis reactions of biomass was extensively discussed in an earlier study (Demirbas, 2000b). Water is formed by dehydration. In pyrolysis reactions, methanol arises from the breakdown of methyl esters and/or ethers from the decomposition of pectinlike plant materials (Goldstein, 1981). Methanol also arises from methoxyl groups of uronic acid (Demirbas and Güllü, 1998). Acetic acid is formed in the thermal decomposition of all three main components of wood. When the yield of acetic acid originating from the cellulose, hemicelluloses, and lignin is taken into account, the total is considerably less than the yield from the wood itself (Wenzl *et al.*, 1970). Acetic acid comes from the elimination of acetyl groups originally linked to the xylose unit.

Furfural is formed by dehydration of the xylose unit. Quantitatively, 1-hydroxy-2-propanone and 1-hydroxy-2-butanone present high concentrations in liquid products. These two alcohols are partly esterified by acetic acid. In conventional slow pyrolysis, these two products are not found in so great a quantity because of their low stability (Beaumont, 1985).

If wood is completely pyrolyzed, the resulting products are about what would be expected by pyrolyzing the three major components separately. The hemicelluloses would break down first, at temperatures of 470 to 530 K. Cellulose follows in the temperature range 510 to 620 K, with lignin being the last component to pyrolyze at temperatures of 550 to 770 K. In one study, a wide spectrum of organic substances was contained in the pyrolytic liquid fractions given in the literature (Beaumont, 1985). Degradation of xylan yields eight main products: water; methanol; formic, acetic, and propionic acids; 1-hydroxy-2-propanone; 1-hydroxy-2-butanone; and 2-furfuraldeyde. The methoxy phenol concentration decreased with increasing temperature, while phenols and alkylated phenols increased. The formation of both methoxy phenol and acetic acid was possibly a result of the Diels-Alder cycloaddition of a conjugated diene and unsaturated furanone or butyrolactone.

Timell (1967) described the chemical structure of xylan as 4-methyl-3-acetylglucoronoxylan. It has been reported that the first runs in the pyrolysis of pyroligneous acids consist of about 50% methanol, 18% acetone, 7% esters, 6% aldehydes, 0.5% ethyl alcohol, 18.5% water, and small amounts of furfural (Demirbas, 2000b). Pyroligneous acids disappear in high-temperature pyrolysis.

The composition of water-soluble products was not ascertained, but it has been reported to be composed of hydrolysis and oxidation products of glucose such as acetic acid, acetone, simple alcohols, aldehydes, sugars, *etc.* (Sasaki *et al.*, 1998).

Pyroligneous acids disappear in high-temperature pyrolysis. Levoglucosan is also sensitive to heat and decomposes into acetic acid, acetone, phenols, and water. Methanol arises from the methoxyl groups of aronic acid (Demirbas, 2000b).

1.3.1.4 Gasification of Biomass

Gasification is a form of pyrolysis carried out in the presence of a small quantity of oxygen at high temperatures in order to optimize gas production. The resulting gas, known as producer gas, is a mixture of carbon monoxide, hydrogen, and methane, together with carbon dioxide and nitrogen. The gas is more versatile than the original solid biomass (usually wood or charcoal): it can be burned to produce process heat and steam or used in gas turbines to produce electricity.

Biomass gasification technologies are expected to be an important part of the effort to meet these goals of expanding the use of biomass. Gasification technologies provide the opportunity to convert renewable biomass feedstocks into clean fuel gases or synthesis gases. Biomass gasification is the latest generation of biomass energy conversion processes and is being used to improve the efficiency and reduce the investment costs of biomass electricity generation through the use of gas turbine technology. High efficiencies (up to approx. 50%) are achievable using combined-cycle gas turbine systems, where waste gases from the gas turbine are recovered to produce steam for use in a steam turbine. Economic studies show that biomass suffocation plants can be as economical as conventional coal-fired plants (Badin and Kirschner, 1998).

Commercial gasifiers are available in a range of sizes and types and run on a variety of fuels, including wood, charcoal, coconut shells, and rice husks. Power output is determined by the economic supply of biomass, which is limited to 80 MW in most regions (Overend, 1998).

Various gasification technologies include gasifiers where the biomass is introduced at the top of the reactor and the gasifying medium is either directed cocurrently (downdraft) or countercurrently up through the packed bed (updraft). Other gasifier designs incorporate circulating or bubbling fluidized beds. Tar yields can range from 0.1% (downdraft) to 20% (updraft) or greater in the product gases.

The process of synthetic fuels (synfuels) from biomass will lower energy costs, improve waste management, and reduce harmful emissions. This triple assault on plant operating challenges is a proprietary technology that gasifies biomass by reacting it with steam at high temperatures to form a clean-burning synthesis gas (syngas: $CO + H_2$). The molecules in the biomass (primarily carbon, hydrogen, and oxygen) and the molecules in the steam (hydrogen and oxygen) reorganize to form this syngas.

1.3.1.5 Combustion of Biomass

Biomass combustion is a series of chemical reactions by which carbon is oxidized to carbon dioxide and hydrogen is oxidized to water. Oxygen deficiency leads to incomplete combustion and the formation of many products of incomplete combustion. Excess air cools the system. The air requirements depend on the chemical and physical characteristics of the fuel. The combustion of biomass relates to the fuel burn rate, the combustion products, the required excess air for complete combustion, and the fire temperatures.

Characteristics influencing combustion are (a) particle size and specific gravity, (b) ash content, (c) moisture content, (d) extractive content, (f) element (C, H, O, N) content, and (g) structural constituent (cellulose, hemicelluloses, lignin) content.

The particle size of biomass should be as much as 0.6 cm, sometimes more, in a profitable combustion process. Biomass is much less dense and has significantly higher aspect ratios than coal. It is also much more difficult to reduce to small sizes.

Ash or inorganic materials in plants depend on the type of plant and soil contamination in which the plant grows. On average wood contains about 0.5% ash. Ash contents of hard and soft woods are about 0.5 and 0.4%, respectively. Insoluble compounds act as a heat sink in the same way as moisture, lowering combustion efficiency, but soluble ionic compounds can have a catalytic effect on the pyrolysis and combustion of fuel. The composition of mineral matter can vary between and within each biomass sample.

Moisture in biomass generally decreases its heating value. Moisture in biomass is stored in spaces within the dead cells and within the cell walls. When the fuel is dried, the stored moisture equilibrates with the ambient relative humidity. Equilibrium is usually about 20% in air-dried fuel.

The moisture percentage of wood species varies from 41.27 to 70.20%. The heating value of a wood fuel decreases as the moisture content of the wood increases. Moisture content varies from one tree part to another. It is often lowest in the stem and increases toward the roots and the crown. The presence of water in biomass influences its behavior during pyrolysis and affects the physical properties and quality of the pyrolysis liquid.

Again the heat content, which is a very important factor affecting the use of any material as a fuel, is affected by the proportion of combustible organic components (called extractives) present in it. The HHVs of the extractive-free plant parts have been found to be lower than those of the unextracted parts, which indicates a likely positive contribution of extractives to the increase of HHV. Extractive content is an important parameter directly affecting the heating value. A high extractive content of a plant part makes it desirable as fuel.

Both the chemical and the physical composition of a fuel are important determining factors in the characteristics of combustion. Biomass can be analyzed by breaking it down into structural components (known as proximate analysis) or chemical elements (known as ultimate analysis). The heat content is related to the oxidation state of the natural fuels in which carbon atoms generally dominate and overshadow small variations in hydrogen content.

1.3.1.6 Heating Value of Biomass

The higher heating values (HHVs) or gross heat of combustion includes the latent heat of the water vapor products of combustion because the water vapor is allowed to condense to liquid water. In one study, the HHV (in MJ/kg) of the biomass fuels as a function of fixed carbon (FC, wt.%) was calculated from Eq. (1.1) (Demirbas, 1997):

$$\mathrm{HHV} = 0.196(\mathrm{FC}) + 14.119. \tag{1.1}$$

In earlier work (Demirbas *et al.*, 1997), formulae were also developed for estimating the HHVs of fuels from different lignocellulosic materials, vegetable oils, and diesel fuels using their chemical analysis data. For biomass fuels such as coal, the HHV was calculated using the modified Dulong's formula (Perry and Chilton, 1973; Demirbas *et al.*, 1997) as a function of the carbon, hydrogen, oxygen, and nitrogen contents from Eq. (1.2):

$$\mathrm{HHV} = 0.335(\mathrm{CC}) + 1.423(\mathrm{HC}) - 0.154(\mathrm{OC}) - 0.145(\mathrm{NC}), \tag{1.2}$$

where CC is carbon content (wt.%), HC hydrogen content (wt.%), OC oxygen content (wt.%), and NC nitrogen content (wt.%)

The heat content is related to the oxidation state of the natural fuels in which carbon atoms generally dominate and overshadow small variations of hydrogen content. On the basis of literature values for different species of wood, Tillman (1978) also found a linear relationship between HHV and carbon content.

The HHVs of extractive-free samples reflect the HHV of lignin relative to cellulose and hemicelluloses. It was reported (Demirbas, 2001) that cellulose and hemicelluloses (holocellulose) have a HHV of 18.60 kJg^{-1}, whereas lignin has a HHV of 23.26 to 26.58 kJg^{-1}. As discussed by Baker (1982), HHVs reported for a given species reflect only the samples tested and not the entire population of the species. The HHV of a lignocellulosic fuel is a function of its lignin content. In general, the HHV of lignocellulosic fuels increases with increases in their lignin content and the HHV is highly correlated with lignin content. For the model including the lignin content, the regression equation is

$$HHV = 0.0889(LC) + 16.8218, \tag{1.3}$$

where LC is the lignin content (wt.% daf and extractive-free basis).

1.3.2 Hydropower

The water in rivers and streams can be captured and turned into hydropower, also called hydroelectric power. Large-scale hydropower provides about one quarter of the world's total electricity supply, virtually all of Norway's electricity, and more than 40% of the electricity used in developing countries. The technically usable world potential of large-scale hydro is estimated to be over 2200 GW, of which only about 25% is currently exploited.

There are two small-scale hydropower systems: microhydropower systems (MHP), with capacities below 100 kW, and small hydropower systems (SHP), with capacity between 101 kW and 1 MW. Large-scale hydropower supplies 20% of global electricity. In developing countries, considerable potential still exists, but large hydropower projects may face financial, environmental, and social constraints (UNDP, 2000).

The two small-scale hydropower systems being discussed in this section are sites with capacities below 100 kW (referred to as microhydropower systems) and sites with capacity between 101 kW and 1 MW (referred to as small hydropower systems). Microhydropower (MHP) systems that use cross flow turbines and pelton wheels can provide both direct mechanical energy (for crop processing) and electrical energy. However, due to design constraints, turbines up to a capacity of 30 kW are suitable for extracting mechanical energy. Of the total installed capacity of about 12 MW of MHP systems, half is used solely for crop processing. The most popular of the MHP systems is the peltric set, which is an integrated pelton turbine and electricity-generation unit with an average capacity of 1 kW. MHP systems are sometimes described as those having capacities below 100 kW, minihydropower plants are those ranging from 100 to 1,000 kW, and small hydropower (SHP) plants are those that produce from 1 to 30 MW.

Dams are individually unique structures, and dam construction represents the largest structures in terms of basic infrastructure in all nations (Novak *et al.*, 1996). Today, worldwide nearly 500,000 km^2 of land are inundated by reservoirs capable of storing 6,000 km^3 of water. As a result of this distribution of fresh water in reservoirs, small but measurable changes have occurred around the world. The total insalled capacity of hydropower is 640,000 MW (26% of the theoretical potential) generating an estimated 2,380 TWh/year in the world and producing nearly 20% of the world's total supply of electricity. 27,900 MW of the total hydropower is at small-scale sites, generating 115 TWh/year (Penche, 1998; Gleick, 1999; Demirbas, 2002). The NAFTA countries are now the biggest producers, along with Latin America and EU/EFTA regions, but it is estimated that Asia will be generating more hydroelectricty than NAFTA countries by the end of the next decade.

There is no universal consensus on the definition of small hydropower. Some countries of the European Union such as Portugal, Spain, Ireland, Greece, and Belgium accept 10 MW as the upper limit for installed capacity. In Italy the limit is 3 MW, in France 8 MW, in the UK 5 MW, in Canada 20 to 25 MW, and in the USA 30 MW; however, a value of up to 10 MW total capacity is becoming generally accepted as small hydropower in the rest of the world. If total installed capacity of any hydropower system is greater than 10 MW, it is generally accepted as a large hydropower system (Cunningham and Atkinson, 1998; Adiguzel and Tutus, 2002; UNIDO, 2003; Kueny, 2003; IASH, 2004; ISHA, 2004). Small hydro can be further subdivided into minihydro, usually defined as <500 kW, and microhydro, which is <100 kW. The definition of microhydro or small-scale hydro varies in different countries. SHP is one of the most valuable energy sources for the electrification of rural communities. Small hydroelectricity growth serves to decrease the gap in decentralized production between private-sector and municipal activity production. SHP systems supply the energy from flowing or running water and convert it into electrical energy. The potential for SHP systems depends on the availability of water flow where the resource exists. If a well-designed SHP system is established somewhere, it can fit in with its surroundings and will have minimal negative impact on the environment. SHP systems allow for self-sufficiency by using scarce natural water resources. These systems supply low-cost energy that is being used in many developing countries in the world (Romas and De Almedia, 1999; Romas and De Almedia, 2000).

1.3.3 Geothermal Energy

As an energy source, geothermal energy has come of age. Geothermal energy for electricity generation has been produced commercially since 1913, and for four decades on the scale of hundreds of megawatts both for electricity generation and direct use. Use of geothermal energy has increased rapidly during the last three decades. In 2000, geothermal resources were identified in over 80 countries, and there are quantified records of geothermal use in 58 countries (Fridleifsson, 2001).

Geothermal energy is clean, cheap, and renewable and can be utilized in various forms such as space heating and domestic hot water supply, CO_2 and dry-ice production processes, heat pumps, greenhouse heating, swimming and balneology (therapeutic baths), industrial processes, and electricity generation. The main types of direct use are bathing, swimming and balneology (42%), space heating (35%), greenhouses (9%), fish farming (6%), and industry (6%) (Fridleifsson, 2001). Geothermal energy can be utilized in various forms such as electricity generation, direct use, space heating, heat pumps, greenhouse heating, and industrial usage. Electricity is produced with geothermal steam in 21 countries spread over all continents. Low-temperature geothermal energy is exploited in many countries to generate heat, with an estimated capacity of about 10,000 MW thermal.

In the Tuscany region of Italy, a geothermal plant has been operating since the early 1900s. There are also geothermal power stations in the USA, New Zealand, and Iceland. In Southampton (UK) there is a district heating scheme based on geothermal energy. Hot water is pumped up from about 1,800 m below ground. Direct application of geothermal energy can involve a wide variety of end uses, such as space heating and cooling, industry, greenhouses, fish farming, and health spas. It uses mostly existing technology and straightforward engineering. The technology, reliability, economics, and environmental acceptability of the direct use of geothermal energy have been demonstrated throughout the world.

1.3.4 Wind Energy

Renewable energy from wind has been used for centuries to power windmills to mill wheat or pump water. More recently large wind turbines have been designed that are used to generate electricity. This source of energy is non-polluting and freely available in many areas. Wind turbines are becoming more efficient, and the cost of the electricity they generate is falling.

There are wind farms around the world. Because the UK is on the edge of the Atlantic Ocean, it has one of the best wind sources in Europe. Offshore wind farms in coastal waters are being developed because winds are often stronger blowing across the sea. Turbines can produce between 500 kW and 1 MW of electricity. Production of wind-generated electricity has risen from practically zero in the early 1980s to more than 7.5 TWh per year in 1995. Cumulative generating capacity worldwide topped 6,500 MW in late 1997 (Garg and Datta, 1998). Figure 1.8 shows the growth in world wind turbine installed capacity.

Wind energy is a significant resource; it is safe, clean, and abundant and is an indigenous supply permanently available in virtually every nation in the world. Today there are wind farms around the world. Globally, wind power generation

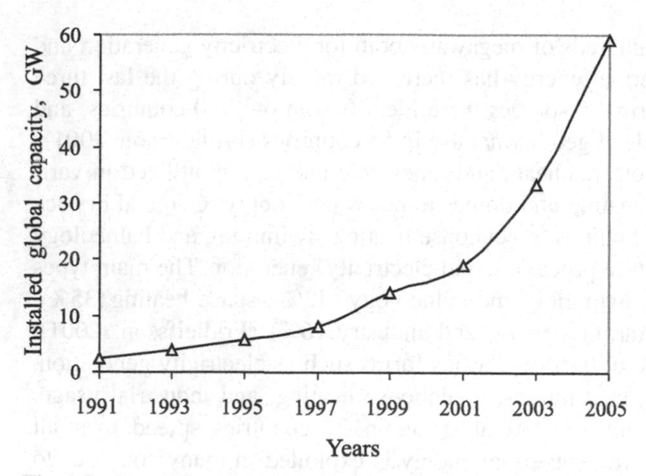

Fig. 1.8 Growth in world wind turbine installed capacity

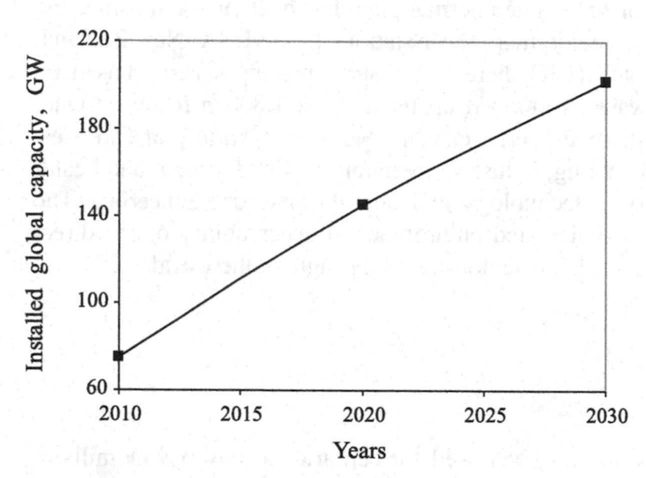

Fig. 1.9 Growth scenarios for global installed wind power

more than quadrupled between 1999 and 2005. Wind energy is abundant, renewable, widely distributed, and clean and mitigates the greenhouse effect if it is used to replace fossil-fuel-derived electricity. Wind energy has limitations based on geography and climate, plus there may be political or environmental problems (*e.g.*, dead birds) with installing turbines (Garg and Datta, 1998). On the other hand, wind can cause air pollution by degradation and distribution of pieces of pollutants such as waste paper, straw, *etc*. Figure 1.9 shows the growth scenarios for global installed wind power (IEA, 2006).

An advantage of wind turbines over some forms of renewable energy is that they can produce electricity whenever the wind blows (at night and also during the day). In theory, wind systems can produce electricity 24 h a day, unlike PV systems, which cannot generate power at night. However, even in the windiest places, the wind does not blow all the time. So while wind farms do not need batteries for backup storage of electricity, small wind systems do need backup batteries. Wind power in coastal and other windy regions is promising as well. By any measure the power in the wind is no longer an alternative source of energy.

1.3.5 Solar Energy

Energy from the sun comes from processes known as solar heating (SH), solar home heating (SHH), solar dryer (SD), solar cooker (SC), solar water heating (SWH), solar photovoltaic (SPV: converting sunlight directly into electricity), and solar thermal electric power (STEP, where the sun's energy is concentrated to heat water and produce steam, which is used to produce electricity). The major component of any solar energy system is the solar collector. Solar energy collectors are special kinds of heat exchangers that transform solar radiation energy into internal energy.

Solar dryers are used for drying fruits and spices. The three most popular types of SD are the box type, the cabinet type, and the tunnel type. The box type uses direct heat for dehydration. In cabinet-type dryers, air heated by the collector dehydrates the food product, whereas in the tunnel type forced air circulation is used to distribute heat for dehydration. Cabinet- and tunnel-type dryers yield a high quality of dried products, but they are very bulky and costly compared to box-type dryers. Of approx. 800 dryers in use so far, 760 are of the box type (Pokharel, 2003).

Solar energy systems are solar home systems, solar photovoltaic (SPV) systems, solar water heating (SWH) systems, solar dryers, and solar cookers. These systems are installed and managed by a household or a small community. A solar home system is a PV system with a maximum capacity of 40 W.

One of the most abundant energy sources is sunlight. Today, solar energy's contribution to the world's total primary energy supply is tiny, less than 1% (Ramachandra, 2007). Photovoltaic (PV) systems, other than solar home heating systems, are used for communication, water pumping for drinking and irrigation, and electricity generation. The total installed capacity of such systems is estimated at *ca.* 1000 kW. A solar home heating system is a solar PV system with a maximum capacity of 40 W. These systems are installed and managed by a household or a small community (Garg and Datta, 1998).

Like wind power markets, PV markets have seen rapid growth and costs have fallen dramatically. The total installed capacity of such systems is estimated at *ca.* 1000 kW. PV installed capacities are growing at a rate of 30% a year (Demirbas, 2005b). Solar PV systems are a promising future energy source. One of the most significant developments in renewable energy production is in PVs (EWEA, 2005; Reijnders, 2006; IEA, 2004). By 2020, PV is projected to be the largest renewable electricity source with a production of 25.1% of global power generation (EWEA, 2005).

PV systems, along with SHH systems, are used for communication, water pumping for drinking and irrigation, and electricity generation. Like wind power markets, PV markets have seen rapid growth and costs have fallen dramatically. The total installed capacity of such systems is estimated at *ca.* 1000 kW. Solar PVs and grid-connected wind installed capacities are growing at a rate of 30% a year (UNDP, 2000).

1.3.6 Biohydrogen

Hydrogen is not a primary fuel. It must be manufactured from water with either fossil or non-fossil energy sources. Widespread use of hydrogen as an energy source could improve global climate change, energy efficiency, and air quality. Thermochemical conversion processes, such as pyrolysis, gasification, and steam gasification, are available for converting biomass into a more useful form of energy. The yield from steam gasification increases with an increasing water-to-sample ratio. The yields of hydrogen from pyrolysis and steam gasification increase with increasing temperature. A list of selected biomass materials used for hydrogen production is given in Table 1.12. Hydrogen-powered fuel cells are an important enabling technology for the hydrogen future and more efficient alternatives to the combustion of gasoline and other fossil fuels. Hydrogen has the potential to solve two major energy problems: dependence on petroleum and pollution and greenhouse gas emissions.

Hydrogen is currently more expensive than conventional energy sources. There are different technologies presently being developed to produce hydrogen economically from biomass. Hydrogen can be produced by pyrolysis from biomass (Arni, 2004). It can be burned to produce heat or passed through a fuel cell to produce electricity. Biohydrogen technology will play a major role in the future because it can utilize renewable sources of energy (Nath and Das, 2003). Figure 1.10 shows the projected future share of hydrogen in total automotive fuel consumption in the world (Demirbas, 2006b).

Table 1.12 List of selected biomass materials used for hydrogen production

Biomass species	Main conversion process
Bio-nut shell	Steam gasification
Olive husk	Pyrolysis
Tea waste	Pyrolysis
Crop straw	Pyrolysis
Black liquor	Steam gasification
Municipal solid waste	Supercritical water extraction
Crop grain residue	Supercritical fluid extraction
Pulp and paper waste	Microbial fermentation
Petroleum basis plastic waste	Supercritical fluid extraction
Manure slurry	Microbiol fermentation

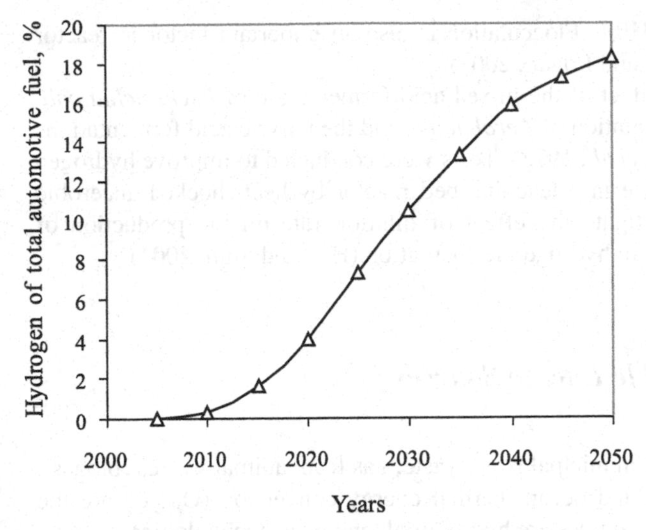

Fig. 1.10 Share of hydrogen in total automotive fuel in the future
Source: Demirbas, 2006b

Biological generation of hydrogen (biohydrogen) technologies provide a wide range of approaches to generating hydrogen, including direct biophotolysis, indirect biophotolysis, photofermentation, and dark fermentation (Levin *et al.*, 2004). Biological hydrogen production processes are found to be more environmentally friendly and less energy intensive as compared to thermochemical and electrochemical processes (Das and Veziroglu, 2001). Researchers have been investigating hydrogen production with anaerobic bacteria since the 1980s (Nandi and Sengupta, 1998; Chang *et al.*, 2002)

There are three types of microorganisms of hydrogen generation: cyanobacteria, anaerobic bacteria, and fermentative bacteria. Cyanobacteria directly decompose water into hydrogen and oxygen in the presence of light energy by photosynthesis. Photosynthetic bacteria use organic substrates like organic acids. Anaerobic bacteria use organic substances as the sole source of electrons and energy, converting them into hydrogen. Biohydrogen can be generated using bacteria such as *Clostridia* by temperature and pH control, reactor hydraulic retention time (HRT), and other factors of the treatment system.

Biological hydrogen can be generated from plants by biophotolysis of water using microalgae (green algae and cyanobacteria), fermentation of organic compounds, and photodecomposition of organic compounds by photosynthetic bacteria. To produce hydrogen by fermentation of biomass, a continuous process using a non-sterile substrate with a readily available mixed microflora is desirable (Hussy *et al.*, 2005). A successful biological conversion of biomass into hydrogen depends strongly on the processing of raw materials to produce feedstock that can be fermented by microorganisms (de Vrije *et al.*, 2002).

Hydrogen production from the bacterial fermentation of sugars has been examined in a variety of reactor systems. Hexose concentration has a greater effect on H_2 yields than HRT. Flocculation is also an important factor in reactor performance (Van Ginkel and Logan, 2005).

Hydrogen gas is a product of the mixed acid fermentation of *Escherichia coli*, the butylene glycol fermentation of *Aerobacter*, and the butyric acid fermentations of *Clostridium* spp. (Aiba *et al.*, 1973). Tests were conducted to improve hydrogen fermentation of food waste in a leaching-bed reactor by heat-shocked anaerobic sludge and also to investigate the effect of dilution rate on the production of hydrogen and metabolites in hydrogen fermentation (Han and Shin, 2004).

1.3.7 Other Renewable Energy Sources

Wave power, tidal power, municipal solid waste, gas from animal wastes (biogas), landfill gas, peat energy, and ocean thermal energy conversion (OTEC) are the other RESs. Water energy sources are hydro, tidal, and wave technologies.

Marine energy sources are current, tidal, ocean thermal energy conversion (OTEC), and wave technologies. The world wave energy capacity is between 200 and 5000 GW and is mostly found in offshore locations (Garg and Datta, 1998). Wave energy converters fixed to the shoreline are likely to be the first to be fully developed and deployed, but waves are typically two to three times more powerful in deep offshore waters than at the shoreline. Wave energy can be harnessed in coastal areas, close to the shore. The first patent for a wave energy device was filed in Paris in 1799, and by 1973 there were 340 British patents for wave energy devices. By comparison to wind and PV, wave energy and tidal stream are very much in their infancy. Currently, around 1 MW of wave energy devices is installed worldwide, mainly from demonstration projects.

Water is a RES that can be used in electricity generation by using its lifting force (buoyant force). It is an important electricity-generating apparatus using gravity, and buoyancy can reduce the costs of power generation and prevent environmental pollution and ecosystem destruction. The hydraulic ram is an attractive solution for electricity generation where a large gravity flow exists. The wave conversion plant using buoyancy chambers is another solution for electricity generation using water-lifting force. There are many reasons why various levels of water-lifting force will be used in the future for electricity generation. Hydraulic ram (hydram) pumps are water-lifting or water-pumping devices that are powered by filling water. Hydram pumps have been used for over two centuries in many parts of the world. They are useful devices that can pump water uphill from a flowing source of water above the source with no power requirement except the force of gravity.

The OTEC is an energy technology that converts solar radiation into electrical power. OTEC systems use the ocean's natural thermal gradient to drive a power-producing cycle. As long as the temperature difference between the warm surface water and the cold deep water is approx. 20 K, an OTEC system can produce a significant amount of power. The oceans are thus a vast renewable resource, with the potential to help us produce billions of watts of electrical power.

Anaerobic digestion in landfills occurs in a series of stages, each of which is characterized by an increase or decrease in specific bacterial populations and the formation and use of certain metabolic products. Landfill gas contains about 50% by volume methane. Producing energy from landfill gas improves local air quality, eliminates a potential explosion hazard, and reduces greenhouse gas emissions into the atmosphere. Hydrogen, produced by passing an electrical current through water, can be used to store solar energy and regenerate it when needed for nighttime energy requirements. Solid waste management practices are collection, recovery, and disposal, together with the results of cost analyses.

Biogas can be obtained from digesting the organic material of municipal solid waste (MSW). The composition of MSW varies by the source of waste; however, in all cases the major constituents of MSW are organic in nature and the organic components account for more than 50% of MSW. The main constituents of landfill gas are methane and carbon dioxide, both of which are major contributors to global warming. The economic exploitation of methane is cost effective starting one year after landfill operations begin.

Biogas can be obtained from several sources. It is obtained from decomposing organic material. It is composed of methane (CH_4), carbon dioxide (CO_2), air, ammonia, carbon monoxide, hydrogen, sulfur gases, nitrogen, and oxygen. Methane is the most important one of its components, particularly for the combustion process in vehicle engines. A typical analysis of raw landfill gas is given in Table 1.13. CH_4 and CO_2 make up *ca.* 90% of the gas volume produced. The main constituents of landfill gas are methane and carbon dioxide, both of which are major contributors to global warming. Because of the widely varying nature of the contents of landfill sites, the constituents of landfill gases vary widely.

The generation of MSW has increased in parallel with rapid industrialization. Approximately 16% of all discarded MSW is incinerated; the remainder is disposed of in landfills. Disposal of MSW in sanitary landfills is usually associated with soil, surface-water, and groundwater contamination when the landfill is not properly constructed. The flow rate and composition of leachate vary from site to site, seasonally at each site and depending on the age of the landfill. The processing of MSW (*i.e.*, landfill, incineration, aerobic composting) has many advantages and

Table 1.13 Typical analysis of raw landfill gas

Component	Chemical formula	Content
Methane	CH_4	40–60 (% by vol.)
Carbon dioxide	CO_2	20–40 (% by vol.)
Nitrogen	N_2	2–20 (% by vol.)
Oxygen	O_2	< 1 (% by vol.)
Heavier hydrocarbons	C_nH_{2n+2}	< 1 (% by vol.)
Hydrogen sulfide	H_2S	40–100 ppm
Complex organics	–	1000–2000 ppm

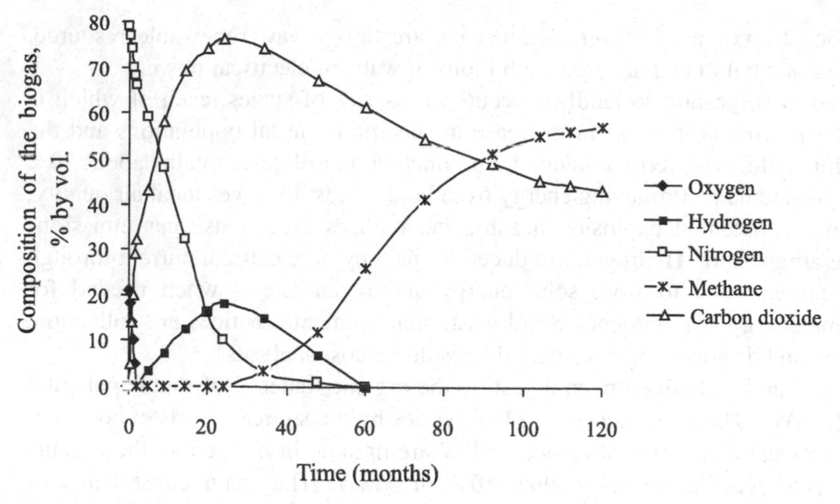

Fig. 1.11 Production of biogas components over time in landfill

limitations. Greenhouse gas emissions can be reduced by the uncontrolled release of methane from improperly disposed organic waste in a large landfill.

The first stage of decomposition, which usually lasts less than a week, is characterized by the removal of oxygen from the waste by aerobic bacteria. In the second stage, which has been termed the anaerobic acid stage, a diverse population of hydrolytic and fermentative bacteria hydrolyzes polymers, such as cellulose, hemicellulose, proteins, and lipids, into soluble sugars, amino acids, long-chain carboxylic acids, and glycerol. The main components of landfill gas are by-products of the decomposition of organic material, usually in the form of domestic waste, by the action of naturally occurring bacteria under anaerobic conditions. Figure 1.11 shows the production of biogas components over time in landfill.

Methods developed for the treatment of landfill leachates can be classified as physical, chemical, and biological; they are usually used in combination to improve treatment efficiency. Biological treatment methods used for leachate treatment can be classified as aerobic, anaerobic, and anoxic processes, which are widely used for the removal of biodegradable compounds (Kargi and Pamukoglu, 2004a). The biological treatment of landfill leachate usually results in low nutrient removals because of high chemical oxygen demand (COD), high ammonium-N content, and the presence of toxic compounds such as heavy metals (Uygur and Kargi, 2004). Landfill leachate obtained from solid waste landfill areas contains high COD and ammonium ions, which results in low COD and ammonium removals by direct biological treatment. Several anaerobic and aerobic treatment systems have been studied in landfill leachate. Leachates contain non-biodegradable substrates that are not removed by biological treatment alone, and an increase of leachate input may cause a reduction in substrate removal (Cecen *et al.*, 2003). Raw landfill leachate was subjected to pretreatment by coagulation-flocculation and air stripping of ammonia before biological treatment (Kargi and Pamokoglu, 2004b). To improve the biological treatability of the leachate, coagulation-flocculation and air stripping of ammonia were used as pretreatment. Natural zeolite and bentonite can be used as a novel landfill liner material (Kayabali, 1997).

References

Adeeb, Z. 2004. Glycerol delignification of poplar wood chips in aqueous medium. Energy Edu Sci Technol 13:81–88.

Adiguzel, F., Tutus, A. 2002. Small hydroelectric power plants in Turkey. In: Proceedings of Hydro 2002, Development, Management Performance, 4–7 November 2002, pp. 283–293.

AER (Annual Energy Review) 1999. 2000. Energy Information Administration, US Department of Energy, Washington, D.C.

Aiba, S., Humphrey, A.E., Milis, N.F. 1973. Biochemical Engineering, 2d edn. Academic, New York.

An, J., Bagnell, L., Cablewski, T., Strauss, C.R., Trainor, R.W. 1997. Applications of high-temperature aqueous media for synthetic organic reactions. J Org Chem 62:2505–2511.

Appel, H.R., Fu, Y.C., Friedman, S., Yavorsky, P.M., Wender, I. 1971. Converting organic wastes to oil. US Burea of Mines Report of Investigation, No. 7560.

Arni, S. 2004. Hydrogen-rich gas production from biomass via thermochemical pathways. Energy Edu Sci Technol 13:47–54.

Babu, B.V., Chaurasia, A.S. 2003. Modeling for pyrolysis of solid particle: kinetics and heat transfer effects. Energy Convers Mgmt 44:2251–2275.

Badin, J., Kirschner, J. 1998. Biomass greens US power production. Renew Energy World 1:40–45

Baker, A.J. 1982. Wood fuel properties and fuel products from woods. In: Proc. Fuelwood, Management and Utilization Seminar, Michigan State University, East Lansing, MI, 9–11 Nov. 1982.

Bala, B.K. 2005. Studies on biodiesels from transformation of vegetable oils for diesel engines. Energy Edu Sci Technol 5:1–45.

Beaumont, O. 1985. Flash pyrolysis products from beech wood. Wood Fiber Sci 17:228–239.

Beckman, D., Boocock, D.G.B. 1983. Liquefaction of wood by rapid hydropyrolysis. Can J Chem Eng 61:80–86.

Boocock, D.G.B., Mackay, D., Lee, P. 1982. Wood liquefaction: extended batch reactions Raney nickel catalyst. Can J Chem Eng 60:802–808.

Cecen, F., Erdincler, A., Kilic, E. 2003. Effect of powdered activated carbon addition on sludge dewaterability and substrate removal in landfill leachate treatment. Adv Environ Res 7: 707–713.

Chang, J.-S., Lee, K.-S., Lin, P.-J. 2002. Biohydrogen production with fixed-bed biorcactors. J Hydrogen Energy 27:1167–1174.

Chornet, E., Overend, R.P. 1985. Fundamentals of thermochemical biomass conversion. Elsevier, New York, pp. 967–1002.

Cuff, D.J., Young, W.J. 1980. US Energy Atlas. Free Press/MacMillan, New York.

Cunningham, P., Atkinson, B. 1998. Micro hydro power in the nineties. http://www.microhydropower.com.

Das, D., Veziroglu, T.N. 2001. Hydrogen production by biological processes: a survey of the literature. Int J Hydrogen Energy 26:13–28

Demirbas, A. 1985. A new method of wood liquefaction. Chim Acta Turc 13:363–368.

Demirbas, A. 1991. Catalytic conversion of residual lignocellulosic materials to an acetone-soluble oil. Sci Technol Int 9:425–433.

Demirbas, A. 1994. Chemicals from forest products by efficient extraction methods. Fuel Sci Technol Int 12:417–431.

Demirbas, A. 1997. Calculation of higher heating values of biomass fuels. Fuel 76:431–434.

Demirbas, A. 1998. Aqueous glycerol delignification of wood chips and ground wood. Bioresour Technol 63:179–185.

Demirbas, A. 2000a. Biomass resources for energy and chemical industry. Energy Edu Sci Technol 5:21–45.

Demirbas, A. 2000b. Mechanisms of liquefaction and pyrolysis reactions of biomass. Energy Convers Mgmt 41:633–646.

Demirbas, A. 2000c. Recent advances in biomass conversion technologies. Energy Edu Sci Technol 6:19–40.

Demirbas, A. 2001. Relationships between lignin contents and heating values of biomass. Energy Convers Mgmt 42:183–188.

Demirbas, A. 2002. Sustainable development of hydropower energy in Turkey. Energy Sources 2:27–40.

Demirbas, A. 2004a. Combustion characteristics of different biomass fuels. Prog Energy Combust Sci 30:219–230.

Demirbas, A. 2004b. Linear equations on thermal degradation products of wood chips in alkaline glycerol. Energy Convers Mgmt 45:983–994.

Demirbas, A. 2005a. Options and trends of thorium fuel utilization in Turkey. Energy Sources 27:597–603.

Demirbas, A. 2005b. Potential applications of renewable energy sources, biomass combustion problems in boiler power systems and combustion related environmental issues. Progress Energy Combust Sci 31:171–192.

Demirbas, A. 2006a. Energy priorities and new energy strategies. Energy Edu Sci Technol 16:53–109.

Demirbas, A. 2006b. Global biofuel strategies. Energy Edu Sci Technol 17:32–63.

Demirbas, A. 2007. The influence of temperature on the yields of compounds existing in bio-oils obtained from biomass samples via pyrolysis. Fuel Proc Technol 88:591–597.

Demirbas, A., Güllü, D. 1998. Acetic acid, methanol and acetone from lignocellosics by pyrolysis. Edu Sci Technol 1:111–115.

Demirbas, A., Gullu D., Caglar A., Akdeniz, F. 1997. Determination of calorific values of fuel from lignocellulosics. Energy Sour 19:765–770.

de Vrije, T., de Haas, G.G., Tan, G.B., Keijsers, E.R.P., Claassen, P.A.M. 2002. Pretreatment of Miscanthus for hydrogen production by Thermotoga elfii. Int J Hydrogen Energy 27: 1381–1390.

Dogru, M., Howarth, C.R., Akay, G., Keskinler, B., Malik, A.A. 2002. Gasification of hazelnut shells in a downdraft gasifier. Energy 27:415–427.

Eager, R.L., Mathews, J.F., Pepper, J.M. 1982. Liquefaction of aspen poplar wood. Can J Chem Eng 60:289–94.

Edinger, R., Kaul, S. 2000. Humankind's detour toward sustainability: past, present, and future of renewable energies and electric power generation. Renew Sustain Energy Rev 4:295–313.

EIA (Energy Information Administration). 2006. Monthly Energy Review, August 2006. http://www.eia.doe.gov/emeu/mer/contents.html.

EIA (Energy Information Administration). 2006. Annual Energy Outlook 2006. U.S. Department of Energy.

EREC (European Renewable Energy Council). 2006. Renewable Energy Scenario by 2040, EREC Statistics, Brussels.

EWEA (European Wind Energy Association). 2005. Report: Large scale integration of wind energy in the European power supply: analysis, issues and recommendations, Brussels.

Fridleifsson, I.B. 2001. Geothermal energy for the benefit of the people. Renew Sustain Energy Rev 5:299–312.

Garg, H.P., Datta, G. 1998. Global status on renewable energy. In: Solar Energy Heating and Cooling Methods in Building, International Workshop: Iran University of Science and Technology, 19–20 May 1998.

Glasser, W.G. 1985. In: R.P. Overand, T.A., L.K. Mudge (eds.) Fundamentals of Thermochemical Biomass Conversion. Elsevier, New York.

Gleick, P.H. 1999. The world's water, the biennial report on freshwater resources. Pacific Institute for Studies in Development, Environment, and Security, Oakland, CA.

Goldstein, I.S. 1981. Organic Chemical from Biomass. CRC Press, Boca Raton, FL, p. 13.

Hall, D.O., Rosillo-Calle, F., de Groot, P. 1992. Biomass energy lessons from case studies in developing countries. Energy Policy 20:62–73.

Han, S.-K., Shin, H.-S. 2004. Biohydrogen production by anaerobic fermentation of food waste. Int J Hydrogen Energy 29:569–577.

Hussy, I., Hawkes, F.R., Dinsdale, R., Hawkes, D.L. 2005. Continuous fermentative hydrogen production from sucrose and sugarbeet. Int J Hydrogen Energy 30:471–483.

Hsu, C.-C., Hisxon, A.N. 1981. C1 to C4 oxygenated compounds by promoted pyrolysis of cellulose. Ind Eng Chem Prod Res Develop 20:109–14.

Hwang, D.W., Kim, H.G., Jang, J.S., Bae, S.W., Ji, S.M., Lee, J.S. 2004. Photocatalytic decomposition of water–methanol solution over metal-doped layered perovskites under visible light irradiation. Catal Today 93–95:845–850.

IASH. 2004. International Association for Small Hydro (www.iash.info/definition.htm)

IEA (International Energy Annual). 2000. Energy Information Administration, US Department of Energy, Washington, D.C.

IEA (International Energy Agency). 2004. Biofuels For Transport: An International Perspective. 9, rue de la Fédération, 75739 Paris, cedex 15, France. http://www.iea.org.

UNIDO. 2003. United Nations Industrial Development Organization. http://www.unido.org.

ISHA. 2004. International Small Hydro Atlas. http://www.small-hydro.com.

Jean-Baptiste, P., Ducroux, R. 2003. Energy policy and climate change. Energy Policy 31: 155–166

Jefferson, M. 2006. Sustainable energy development: performance and prospects. Renew Energy 31:571–582.

Kalogirou, S. A. 2004. Solar thermal collectors and applications. Prog Energy Combust Sci 30:231–295.

Kargi, F., Pamukoglu, M.Y. 2004a. Adsorbent supplemented biological treatment of pre-treated landfill leachate by fed-batch operation. Bioresour Technol 94:285–291.

Kargi, F., Pamukoglu, M.Y. 2004b. Repeated fed-batch biological treatment of pre-treated landfill leachate by powdered activated carbon addition. Enzyme Microb Technol 34: 422–428.

Kayabali, K. 1997. Engineering aspects of a novel landfill liner material: bentonite-amended natural zeolite. Eng Geol 46:105–114.

Kucuk, M.M., Demirbas, A. 1993. Delignification of Ailanthus altissima and Spruce orientalis with glycerol or alkalin glycerol at atmospheric pressure. Cellulose Chem Technol 27: 679–686.

Kueny, J.L. 2003. Objectives for small hydro technology, Part I. Institute National Polytechnique de Grenoble, pp. 1–35, B.P. 53–38041- Grenoble Cedex 9-France. http://www.small-hydro.com.

Kuhlmann, B., Arnett, E.M., Siskin, M. 1994. Classical organic reactions in pure superheated water. J Organ Chem 59:3098–3101.

Levin, D.B., Pitt, L., Love, M. 2004. Biohydrogen production: prospectsand limitationsto practical application. Int J Hydrogen Energy 29:173–185.

McGowan, F. 1991. Controlling the greenhouse effect: the role of renewables. Energy Policy 19:110–118.

Minowa, T., Ogi, T., Dote, Y.,Yokoyama, S. 1994. Effect of lignin content on direct liquefaction of bark. Int Chem Eng 34:428–430.

Mohan, D., Pittman, Jr., C.U., Steele, P.H. 2006. Pyrolysis of wood/biomass for bio-oil: a critical review. Energy Fuels 20:848–889.
Nandi, R., Sengupta, S. 1998. Microbial production of hydrogen—an overview. Crit Rev Microbiol 24:61–84.
Nath, K., Das, D. 2003. Hydrogen from biomass. Curr Sci 85:265–271.
Novak, P., Moffat, A.I.B., Nalluri, C., Narayanan, R. 1996. Hydraulic structures, 2nd edn. E & FN Spon, an imprint of Chapman & Hall, London.
Ogi, T., Yokoyama, S., Koguchi, K. 1985. Direct liquefaction of wood by alkali and alkaline earth salt in an aqueous phase. Chem Lett 8:1199–200.
Ogi, T., Yokoyama, S. 1983. Liquid fuel production from woody biomass by direct liquefaction. Sekiyu Gakkaishi 36:73–84.
Overend, R.P. 1998. Biomass gasification: a growing business. Renew Energy World 1:59–63.
Ozcimen, D., Karaosmanoglu, F. 2004. Production and characterization of bio-oil and biochar from rapeseed cake. Renew Energy 29:779–787.
Penche, C. 1998. Layman's guidebook on how to develop a small hdro site, ESHA, European Small Hydropower Association, Directorate General for Energy (DG XVII).
Perry, S.W., Chilton, C.N. 1973. Chemical Engineers' Handbook, 5th edn. McGraw-Hill, New York.
Phillips, V.D., Kinoshita, C.M., Neill, D.R., Takahashi, P.K. 1990. Thermochemical production of methanol from biomass in Hawaii. Appl Energy 35:167–175.
Pokharel, S. 2003. Promotional issues on alternative energy technologies in Nepal. Energy Policy 31:307–318.
Qu, Y., Wei, X., Zhong, C. 2003. Experimental study on the direct liquefaction of Cunninghamia lanceolata in water. Energy 28:597–606.
Ramachandra, T.V. 2007. Solar energy potential assessment using GIS. Energy Edu Sci Technol 18:101–114.
Ramage, J., Scurlock, J. 1996. Biomass. In: G. Boyle (ed.) Renewable Energy-Power for a Sustainable Future. Oxford University Press, Oxford.
Reijnders, L. 2006. Conditions for the sustainability of biomass based fuel use. Energy Policy 34:863–876.
Romas, H., De Almedia, A.B. 1999. Small hydropower schemes as an important Renewable Energy Sources, Hidroenergia 99, 1–13 October 1999, Vienna, Austria.
Romas, H., De Almedia, A.B. 2000. Small hydro as one of the oldest renewable energy sources. In: Water Power and Dam Construction, Small Hydro, Lisbon, Portugal, 8–12 May 2000.
Rowell, R.M., Hokanson, A.E. 1979. Methanol from Wood: A Critical Assessment. In: Sarkanen, K.V., Tillman, D.A. (eds.) Progress in Biomass Conversion, vol. 1. Academic, New York.
Rydholm, S.A. 1965. Pulping Processes. Interscience, New York
Sasaki, M., Kabyemela, B.M., Malaluan, R.M., Hirose, S., Takeda, N., Adschiri, T., Arai, K. 1998. Cellulose hydrolysis in subcritical and supercritical water. J Supercrit Fluids 13:261–268.
Schobert, H.H., Song, C. 2002. Chemicals and materials from coal in the 21st century. Fuel 81:15–32.
Sorensen, H.A. 1983. Energy Conversion Systems. Wiley, New York.
Taner, F., Eratik, A., Ardic, I. 2004. Identification of the compounds in the aqueous phases from liquefaction of lignocellulosics. Fuel Process Technol 86:407–418.
Theander, O. 1985. In: Overand, R.P., Mile, T.A., Mudge, L.K. (eds.) Fundamentals of Thermochemical Biomass Conversion. Elsevier, New York.
Tillman, D.A. 1978. Wood as an Energy Resource. Academic, New York.
Timell, T.E. 1967. Recent progress in the chemistry of wood hemicelluloses. Wood Sci Technol 1:45–70.
UNDP (United Nations Development Programme). 2000. World energy assessment 2000—energy and the challenge of sustainability. New York (ISBN 9211261260).

Uygur, A., Kargi, F. 2004. Biological nutrient removal from pre-treated landfill leachate in a sequencing batch reactor. J Environ Mgmt 71:9–14.

Van Ginkel, S.W., Logan, B. 2005. Increased biological hydrogen production with reduced organic loading. Water Res 39:3819–3826.

Wenzl, H.F.J., Brauns, F.E., Brauns, D.A. 1970. The Chemical Technology of Wood. Academic, New York.

WEC (Word Energy Council). 2002. Survey of Energy Sources. New York.

Yazıcı N., Demirbas A. 2001. Turkey's natural gas necessity and Turkey's natural gas consumption. Energy Sources 23:801–808.

Chapter 2
Biofuels

2.1 Introduction to Biofuels

Known petroleum reserves are limited and will eventually run out. Various studies put the date of the global peak in oil production between 1996 and 2035. Biomass energy technologies use waste or plant matter to produce energy with a lower level of greenhouse gas emissions than fossil fuel sources (Sheehan *et al.*, 1998). In developed countries there is a growing trend toward employing modern technologies and efficient bioenergy conversion using a range of biofuels, which are becoming cost competitive with fossil fuels (Puhan *et al.*, 2005). The biofuel economy will grow rapidly during the 21st century. The biofuel economy, and its associated biorefineries, will be shaped by many of the same forces that shaped the development of the hydrocarbon economy and its refineries over the past century. President Bush spoke in his January 31, 2006 State of the Union address of producing biofuels by 2012 using "woodchips, stalks, and switchgrass" as the source of cellulosic biomass. These represent both existing and potential biomass resources. Due to the environmental merits of biofuel, its share in the automotive fuel market will grow rapidly in the next decade.

Various scenarios have put forward estimates of biofuel from biomass sources in the future energy system. In the most biomass-intensive scenario, by 2050 modernized biomass energy will contribute about one half of the total energy demand in developing countries (IPCC, 1997). The biomass-intensive future energy supply scenario includes 385 million ha of biomass energy plantations globally in 2050, with three quarters of this area established in developing countries (Kartha and Larson, 2000). The availability of biofuel resources is important for the electricity, heat, and liquid fuel market. There are two global biomass-based liquid transportation fuels that might replace gasoline and diesel fuel. These are bioethanol and biodiesel. Transport is one of the main energy-consuming sectors. It is assumed that biodiesel will be used as a fossil diesel replacement and that bioethanol will be used as a gasoline replacement. Biomass-based energy sources for heat, electricity, and transportation fuels are potentially carbon dioxide neutral and recycle the same carbon atoms. Due to the widespread availability of biomass resources, biomass-based fuel technology can potentially employ more people than fossil-fuel-based technology (Demirbas, 2006a). Demand for energy is increasing every day due to the rapid growth of population and urbanization. As the major conventional energy sources like coal, petroleum, and natural gas are gradually depleted, biomass is emerging as one of the promising environmentally friendly renewable energy options.

The term biofuel refers to liquid or gaseous fuels for the transport sector that are predominantly produced from biomass. It is generally held that biofuels offer many benefits, including sustainability, reduction of greenhouse gas emissions, and security of supply (Reijnders, 2006). A variety of fuels can be produced from biomass resources including liquid fuels, such as ethanol, methanol, biodiesel, and Fischer-Tropsch diesel, and gaseous fuels, such as hydrogen and methane. Biofuels are primarily used in vehicles but can also be used in engines or fuel cells for electricity generation.

There are several reasons why biofuels are considered relevant technologies by both developing and industrialized countries (Demirbas, 2006a). They include energy security, environmental concerns, foreign exchange savings, and socioeconomic issues related to the rural sector. Due to its environmental merits, the share of biofuel in the automotive fuel market will grow fast in the next decade (Kim and Dale, 2005; Demirbas and Balat, 2006). The advantages of biofuels are the following: (a) they are easily available from common biomass sources, (b) carbon dioxide cycle occurs in combustion, (c) they are very environmentally friendly, and (d) they are biodegradable and contribute to sustainability (Puppan, 2002).

Various scenarios have led to the conclusion that biofuels will be in widespread use in the future energy system. The scenarios are to facilitate the transition from the hydrocarbon economy to the carbohydrate economy by using biomass to produce bioethanol and biomethanol as replacements for traditional oil-based fuels and feedstocks. The biofuel scenario produces equivalent rates of growth in GDP and per-capita affluence, reduces fossil energy intensities of GDP, and reduces oil imports. Each scenario has advantages whether in terms of rates of growth in GDP, reductions in carbon dioxide emissions, the energy ratio of the production process, the direct creation of jobs, or the area of biomass plantation required to make the production system feasible (Demirbas, 2006a).

The biggest difference between biofuels and petroleum feedstocks is oxygen content. Biofuels have oxygen levels of 10 to 45% while petroleum has essentially none, making the chemical properties of biofuels very different from those of petroleum. All have very low sulfur levels and many have low nitrogen levels.

Biomass can be converted into liquid and gaseous fuels through thermochemical and biological methods. Biofuel is a non-polluting, locally available, accessible, sustainable, and reliable fuel obtained from renewable sources (Vasudevan *et al.*, 2005). Liquid biofuels fall into the following categories: (a) vegetable oils and biodiesels, (b) alcohols, and (c) biocrude and synthetic oils. Figure 2.1 shows the sources of the main liquid biofuels for automobiles.

Biomass is an attractive feedstock for three main reasons. First, it is a renewable resource that could be sustainably developed in the future. Second, it appears

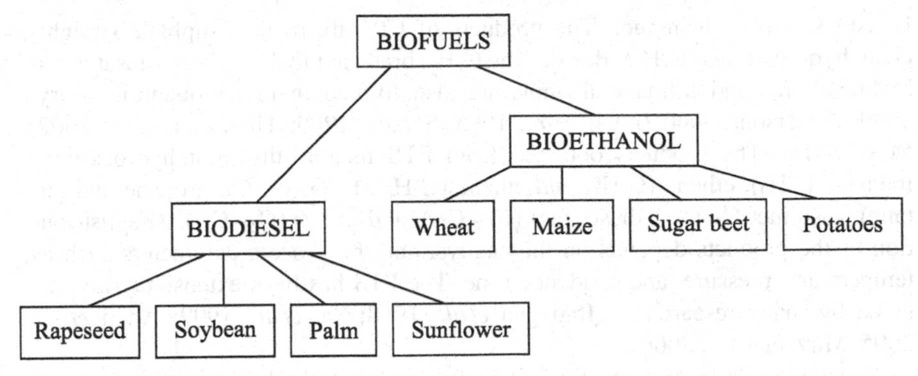

Fig. 2.1 Sources of main liquid biofuels for automobiles

to have formidably positive environmental properties resulting in no net releases of carbon dioxide and very low sulfur content. Third, it appears to have significant economic potential provided that fossil fuel prices increase in the future (Cadenas and Cabezudo, 1998). Lignocellulosic biomethanol has such low emissions because the carbon content of the alcohol is primarily derived from carbon that was sequestered in the growing of the biofeedstock and is only being rereleased into the atmosphere (Difiglio, 1997).

Carbohydrates (hemicelluloses and cellulose) in plant materials can be converted into sugars by hydrolysis. Fermentation is an anaerobic biological process in which sugars are converted into alcohol by the action of microorganisms, usually yeast. The resulting alcohol is bioethanol. The value of any particular type of biomass as feedstock for fermentation depends on the ease with which it can be converted into sugars. Bioethanol is a petrol additive/substitute. It is possible that wood, straw, and even household wastes may be economically converted into bioethanol. Ethanol demand is expected to more than double in the next ten years. For the supply to be available to meet this demand, new technologies must be moved from the laboratories to commercial reality (Bothast and Schlicher, 2005). World ethanol production is about 60% from sugar-crop feedstock.

Anaerobic digestion of biowastes occurs in the absence of air; the resulting gas, called biogas, is a mixture consisting mainly of methane and carbon dioxide. Biogas is a valuable fuel that is produced in digesters filled with feedstock like dung or sewage. The digestion is allowed to continue for a period of ten days to a few weeks (Demirbas, 2006b).

The Fischer–Tropsch synthesis (FTS) produces hydrocarbons of different lengths from a gas mixture of H_2 and CO (syngas) resulting from biomass gasification called bio-syngas. The fundamental reactions of synthesis gas chemistry are methanol synthesis, FTS, oxosynthesis (hydroformylation), and methane synthesis (Prins *et al.*, 2004). The FTS process is capable of producing liquid hydrocarbon fuels from bio-syngas. The large hydrocarbons can be hydrocracked to form mainly diesel of excellent quality. The process for producing liquid fuels from biomass, which integrates biomass gasification with FTS, converts a renewable feedstock into a clean fuel. The products of FTS are mainly aliphatic straight-chain hydrocarbons (C_xH_y). Besides the C_xH_y, branched hydrocarbons, unsaturated hydrocarbons, and primary alcohols are also formed in minor quantities (Dry, 1999; Anderson, 1984; Bukur *et al.*, 1995; Schulz, 1999; Tijmensen *et al.*, 2002; May, 2003). The products obtained from FTS include the light hydrocarbons methane (CH_4), ethene (C_2H_4) and ethane (C_2H_6), LPG (C_3–C_4, propane and butane), gasoline (C_5–C_{12}), diesel fuel (C_{13}–C_{22}), and waxes (C_{23}–C_{33}). The distribution of the products depends on the catalyst and the process parameters such as temperature, pressure, and residence time. The FTS has been extensively investigated by many researchers (Rapagna *et al.*, 1998; Sie *et al.*, 1999; Ahón *et al.*, 2005; Mirzaei *et al.*, 2006).

Vegetable oils from renewable oil seeds can be used when mixed with diesel fuels. Pure vegetable oil, however, cannot be used in direct-injection diesel engines, such as those regularly used in standard tractors, since engine cooking occurs after several hours of use. Conversion of vegetable oils and animal fats into biodiesel has been undergoing further development over the past several years (Prakas, 1998; Madras *et al.*, 2004; Haas *et al.*, 2006; Meher *et al.*, 2006). Biodiesel represents an alternative to petroleum-based diesel fuel. Chemically speaking, biodiesel is a mixture of monoalkyl esters of fatty acids, most often obtained from extracted plant oils and/or collected animal fats. Commonly accepted biodiesel raw materials include the oils from soy, canola, corn, rapeseed, and palm. New plant oils that are under consideration include mustard seed, peanut, sunflower, and cotton seed. The most commonly considered animal fats include those derived from poultry, beef, and pork (Usta *et al.*, 2005).

Ethanol is the most widely used liquid biofuel. It is an alcohol and is fermented from sugars, starches, or cellulosic biomass. Most commercial production of ethanol is from sugar cane or sugar beet, as starches and cellulosic biomass usually requires expensive pretreatment. Ethanol is used as a renewable energy fuel source as well as for the manufacture of cosmetics and pharmaceuticals and also for the production of alcoholic beverages. Ethyl alcohol is not only the oldest synthetic organic chemical used by humans, but it is also one of the most important. In an earlier study (Taherzadeh, 1999), the physiological effects of inhibitors on ethanol from lignocellulosic materials and fermentation strategies were comprehensively investigated.

2.2 Bioethanol

Carbohydrates (hemicelluloses and cellulose) in plant materials can be converted into sugars by hydrolysis. Fermentation is an anaerobic biological process in which sugars are converted into alcohol by the action of microorganisms, usually yeast. The resulting alcohol is ethanol. The value of any particular type of biomass as feedstock for fermentation depends on the ease with which it can be converted into sugars.

Table 2.1 Ethanol production on different continents (billion liters/year)

America	Asia	Europe	Africa	Oceania
22.3	5.7	4.6	0.5	0.2

Cellulose is a remarkable pure organic polymer, consisting solely of units of anhydroglucose held together in a giant straight-chain molecule. Cellulose must be hydrolyzed into glucose before being fermented into ethanol. The conversion efficiencies of cellulose into glucose may depend on the extent to which chemical and mechanical pretreatments are able to structurally and chemically alter the pulp and paper mill wastes. The method of pulping, the type of wood, and the use of recycled pulp and paper products could also influence the accessibility of cellulose to cellulase enzymes (Adeeb, 2004).

Cellulose is insoluble in most solvents and has a low accessibility to acid and enzymatic hydrolysis. Hemicelluloses (arabinoglycuronoxylan and galactoglucomammans) are related to plant gums in composition and occur in much shorter molecule chains than cellulose. Hemicelluloses, which are present in deciduous woods chiefly as pentosans and in coniferous woods almost entirely as hexosanes, undergo thermal decomposition very readily. Hemicelluloses are derived mainly from chains of pentose sugars and act as the cement material holding together the cellulose micelles and fiber (Theander, 1985). Hemicelluloses are largely soluble in alkali and as such are more easily hydrolyzed.

Bioethanol is a fuel derived from renewable sources of feedstock, typically plants such as wheat, sugar beet, corn, straw, and wood. Bioethanol is a petrol additive/substitute. Table 2.1 shows ethanol production on different continents.

Bioethanol can be used as a 5% blend with petrol under EU quality standard EN 228. This blend requires no engine modification and is covered by vehicle warranties. With engine modification, bioethanol can be used at higher levels, for example, E85 (85% bioethanol).

Bioethanol can be produced from a large variety of carbohydrates with a general formula of $(CH_2O)_n$. Fermentation of sucrose is performed using commercial yeast such as *Saccharomyces ceveresiae*. The chemical reaction consists of enzymatic hydrolysis of sucrose followed by fermentation of simple sugars (Berg, 1988). First, invertase enzyme in the yeast catalyzes the hydrolysis of sucrose to convert it into glucose and fructose:

$$\underset{\text{Sucrose}}{C_{12}H_{22}O_{11}} \rightarrow \underset{\text{Glucose}}{C_6H_{12}O_6} + \underset{\text{Fructose}}{C_6H_{12}O_6} \tag{2.1}$$

Second, zymase, another enzyme also present in yeast, converts the glucose and the fructose into ethanol:

$$C_6H_{12}O_6 \rightarrow 2\,C_2H_6OH + 2CO_2. \tag{2.2}$$

Glucoamylase enzyme converts the starch into D-glucose. The enzymatic hydrolysis is then followed by fermentation, distillation, and dehydration to yield

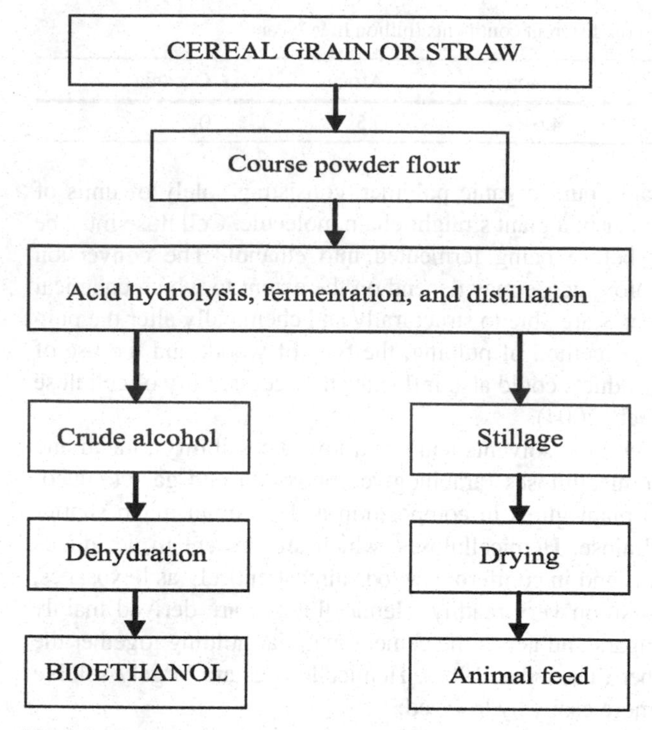

Fig. 2.2 Flow chart for the production of bioethanol from cereal grain or straw

anhydrous bioethanol. Corn (60 to 70% starch) is the dominant feedstock in the starch-to-bioethanol industry worldwide.

Carbohydrates (hemicelluloses and cellulose) in lignocellulosic materials can be converted into bioethanol. The lignocellulose is subjected to delignification, steam explosion, and dilute acid prehydrolysis, which is followed by enzymatic hydrolysis and fermentation into bioethanol (Baltz *et al.*, 1982; Castro *et al.*, 1993; Sokhansanj *et al.*, 2002; Kim and Dale, 2005). A major processing step in an ethanol plant is enzymatic saccharification of cellulose into sugars through treatment by enzymes; this step requires lengthy processing and normally follows a short-term pretreatment step (Kumar *et al.*, 2005). Figure 2.2 shows the flow chart for the production of bioethanol from cereal grain or straw.

Hydrolysis breaks down the hydrogen bonds in the hemicellulose and cellulose fractions into their sugar components: pentoses and hexoses. These sugars can then be fermented into bioethanol. The most commonly applied methods can be classified into two groups: chemical hydrolysis (dilute and concentrated acid hydrolysis) and enzymatic hydrolysis. In chemical hydrolysis, pretreatment and hydrolysis may be carried out in a single step. There are two basic types of acid used in hydrolysis: dilute acid and concentrated acid.

The biggest advantage of dilute acid processes is their fast rate of reaction, which facilitates continuous processing. Since 5-carbon sugars degrade more rapidly than 6-carbon sugars, one way to decrease sugar degradation is to have a two-stage process. The first stage is conducted under mild process conditions to recover the 5-carbon sugars, while the second stage is conducted under harsher conditions to recover the 6-carbon sugars.

Concentrated sulfuric or hydrochloric acid is used for the hydrolysis of lignocellulosic materials. The concentrated acid process uses relatively mild temperatures, and the only pressures involved are those created by pumping materials from vessel to vessel. Reaction times are typically much longer than for dilute acid. This process provides a complete and rapid conversion of cellulose into glucose and hemicelluloses into 5-carbon sugars with little degradation. The critical factors needed to make this process economically viable are to optimize sugar recovery and cost effectively recover the acid for recycling. The solid residue from the first stage is dewatered and soaked in a 30 to 40% concentration of sulfuric acid for 1 to 4 h as a precellulose hydrolysis step. The solution is again dewatered and dried, increasing the acid concentration to about 70%. After reacting in another vessel for 1 to 4 h at low temperatures, the contents are separated to recover the sugar and acid. The sugar/acid solution from the second stage is recycled to the first stage to provide the acid for first-stage hydrolysis (Demirbas, 2006b). The primary advantage of the concentrated acid process is the potential for high sugar recovery efficiency. The acid and sugar are separated via ion exchange, and then the acid is reconcentrated via multiple-effect evaporators.

2.3 Biomethanol

Methanol is one possible replacement for conventional motor fuels. It has been seen as a possible large-volume motor fuel substitute at various times during gasoline shortages. It was often used in the early part of the century to power automobiles before inexpensive gasoline was widely introduced. Synthetically produced methanol was widely used as a motor fuel in Germany during World War II. Methanol is commonly used in biodiesel production for its reactivity. The use of methanol as a motor fuel received attention during the oil crises of the 1970s due to its availability and low cost. Problems occurred early in the development of gasoline-methanol blends. As a result of its low price, some gasoline marketers overblended. Many tests have shown promising results using 85 to 100% by volume methanol as a transportation fuel in automobiles, trucks, and buses.

Methanol, also known as "wood alcohol", is generally easier to find than ethanol. Sustainable methods of methanol production are currently not economically viable. Methanol is produced from synthetic gas or biogas and evaluated as a fuel for internal combustion engines. The production of methanol is a cost-intensive chemical process. Therefore, in current conditions, only waste biomass such as old wood or biowaste is used to produce methanol (Vasudevan *et al.*, 2005).

Methanol is poisonous and burns with an invisible flame. Like ethyl alcohol, it has a high octane rating, and hence an Otto engine is preferable. Most processes require supplemental oxygen for the intermediate conversion of the biomass into a synthesis gas (H_2 + CO). A readily available supply of hydrogen and oxygen, therefore, should improve the overall productivity of biomass-derived methanol (Ouellette *et al.*, 1997).

Before modern production technologies were developed in the 1920s, methanol was obtained from wood as a coproduct of charcoal production and for this reason was commonly known as wood alcohol. Methanol is currently manufactured worldwide by conversion or derived from syngas, natural gas, refinery offgas, coal, or petroleum:

$$2H_2 + CO \rightarrow CH_3OH. \quad (2.3)$$

The chemical composition of syngas from coal and then from natural gas can be identical with the same H_2/CO ratio. A variety of catalysts are capable of causing the conversion, including reduced NiO-based preparations, reduced Cu/ZnO shift preparations, Cu/SiO_2 and Pd/SiO_2, and Pd/ZnO (Takezawa *et al.*, 1987; Iwasa *et al.*, 1993).

Methanol is currently made from natural gas but can also be made using biomass via partial oxidation reactions (Demirbas and Gullu, 1998). Biomass and coal can be considered potential fuels for gasification and further syngas production and methanol synthesis (Takezawa *et al.*, 1987). Adding sufficient hydrogen to the synthesis gas to convert all of the biomass into methanol carbon more than doubles the methanol produced from the same biomass base (Phillips *et al.*, 1990). Waste material can be partially converted into methanol, and the product yield for the conversion process is estimated to be 185 kg of methanol per metric ton of solid waste (Brown *et al.* 1952; Sorensen 1983). Agri-(m)ethanol is at present more expensive than synthesis ethanol from ethylene and methanol from natural gas (Grassi 1999).

Biomass resources can be used to produce methanol. The pyroligneous acid obtained from wood pyrolysis consists of about 50% methanol, acetone, phenols, and water (Demirbas and Gullu, 1998; Gullu and Demirbas, 2001). As a renewable resource, biomass represents a potentially inexhaustible supply of feedstock for methanol production. The composition of bio-syngas from biomass for producing methanol is presented in Table 2.2. Current natural gas feedstocks are so inexpensive that even with tax incentives renewable methanol has not been able to compete economically. Technologies are being developed that may eventually result in the commercial viability of renewable methanol.

Methanol from coal could be a very important source of liquid fuel in the future. The coal is first pulverized and cleaned, then fed to a gasifier bed, where it is reacted with oxygen and steam to produce syngas. Once these steps have been taken, the production process is much the same as with other feedstocks with some variations in the catalyst used and the design of the converter vessel in which the reaction is carried out. Methanol is made using synthesis gas (syngas) with hydrogen and carbon monoxide in a 2-to-1 ratio (Table 2.2). The syngas is transformed into methanol in a fixed-catalyst-bed reactor. Coal-derived methanol has many preferable properties: is is free of sulfur and other impurities, it could replace

Table 2.2 Composition of bio-syngas from biomass gasification

Constituents	% by volume (dry and nitrogen free)
Carbon monoxide (CO)	28–36
Hydrogen (H_2)	22–32
Carbon dioxide (CO_2)	21–30
Methane (CH_4)	8–11
Ethene (C_2H_4)	2–4

petroleum in transportation or be used as a peaking fuel in combustion turbines, or it could serve as a source of hydrogen for fuel cells. The technology for making methanol from natural gas is already in place and requires only efficiency improvements and scaling up to make methanol an economically viable alternative transportation fuel (Demirbas, 2000).

In recent years, a growing interest has been observed in the application of methanol as an alternative liquid fuel that can be used directly for powering Otto engines or fuel cells (Chmielniak and Sciazko, 2003). The feasibility of achieving the conversion has been demonstrated in a large-scale system in which a product gas is initially produced by pyrolysis and gasification of a carbonaceous matter. Syngas from biomass is altered by catalyst under high pressure and temperature to form methanol. This method will produce 100 gallons of methanol per ton of feed material (Rowell and Hokanson, 1979).

The gases produced can be steam reformed to produce hydrogen and followed by water-gas-shift reaction to further enhance hydrogen production. When the moisture content of biomass is higher than 35%, it can be gasified with supercritical water (Hao and Guo, 2002). Supercritical water gasification is a promising process to gasify biomass with high moisture content due to a high gasification ratio (100% achievable) and high hydrogen volumetric ratio (50% achievable) (Yoshida *et al.*, 2004; Matsumura and Minowa, 2004). Hydrogen produced by biomass gasification was reported to be comparable to that by natural gas reforming (Bowen *et al.*, 2003). The process is more advantageous than fossil fuel reforming due to the environmental benefits. It is expected that biomass thermochemical conversion will emerge as one of the most economical large-scale renewable hydrogen technologies.

The strategy is based on producing hydrogen from biomass pyrolysis using a coproduct strategy to reduce the cost of hydrogen and it was concluded that only this strategy could compete with the cost of commercial hydrocarbon-based technologies (Wang *et al.*, 1998). This strategy will demonstrate how hydrogen and biofuels are economically feasible and can foster the development of rural areas when practiced on a larger scale. The process of converting biomass into activated carbon is an alternative route to producing hydrogen with a valuable coproduct that is practiced commercially (Demirbas, 1999).

The simultaneous production of biomethanol (obtained by the hydrogenation of CO_2 developed during the fermentation of sugar juice), in parallel with the production of bioethanol, appears economically attractive in locations where hydroelectricity is available at very low cost (~US$0.01 per Kwh) and where lignocellulosic residues are available as surpluses.

The gas is converted into methanol in a conventional steam-reforming/water-gas-shift reaction followed by high-pressure catalytic methanol synthesis:

$$CH_4 + H_2O \rightarrow CO + 3H_2, \tag{2.4}$$

$$CO + H_2O \rightarrow CO_2 + H_2. \tag{2.5}$$

Eqs. (2.4) and (2.5) are called gasification/shift reactions.

$$CO + 2H_2 \rightarrow CH_3OH \tag{2.6}$$

or

$$CO_2 + 3H_2 \rightarrow CH_3OH + H_2O \tag{2.7}$$

Eqs. (2.6) and (2.7) are methanol synthesis reactions. Figure 2.3 shows production of biomethanol from carbohydrates by gasification and partial oxidation with O_2 and H_2O.

The energy value of residues generated worldwide in agriculture and the forest-products industry amounts to more than one third of the total commercial primary

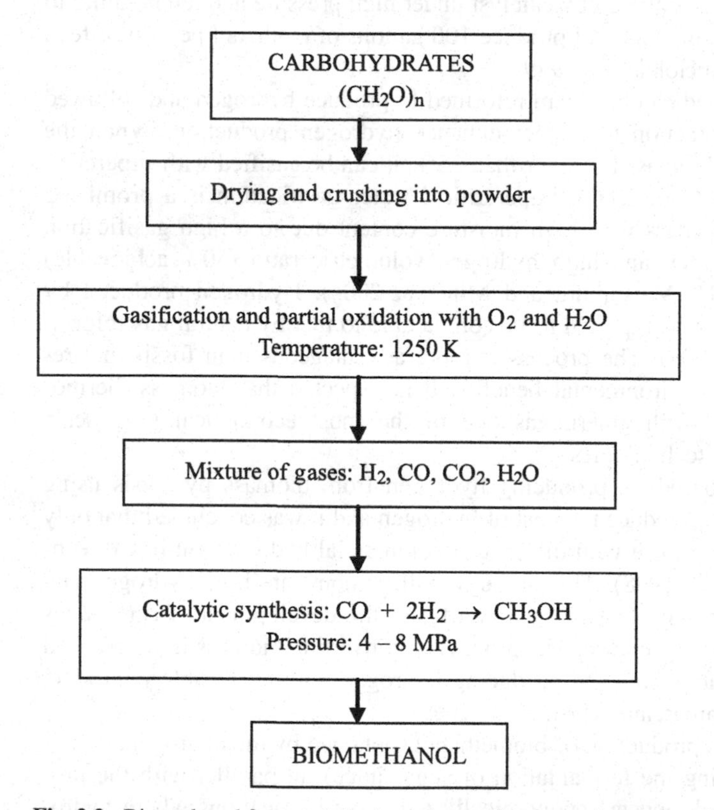

Fig. 2.3 Biomethanol from carbohydrates by gasification and partial oxidation with O_2 and H_2O

Table 2.3 Main production facilities of methanol and biomethanol

Methanol	Biomethanol
Catalytic synthesis from CO and H_2	Catalytic synthesis from CO and H_2
Natural gas	Distillation of liquid from wood pyrolysis
Petroleum gas	Gaseous products from biomass gasification
Distillation of liquid from coal pyrolysis	Synthetic gas from biomass and coal

energy use at present (Hall *et al.*, 1993). Bioenergy supplies can be divided into two broad categories: (a) organic municipal waste and residues from the food and materials sectors and (b) dedicated energy-crop plantations. Bioenergy from biomass, both residues and energy crops, can be converted into modern energy carriers such as hydrogen, methanol, ethanol, or electricity (Azar *et al.*, 2003).

Methanol can be produced from biomass. Thus the choice of fuel in the transportation sector is to some extent determined by the availability of biomass. As regards the difference between hydrogen and methanol production costs, the conversion of natural gas, biomass, and coal into hydrogen is generally more energy efficient and less expensive than conversion into methanol (Azar *et al.*, 2003). The main production facilities of methanol and biomethanol are given in Table 2.3.

2.4 Biohydrogen from Biomass by Steam Reforming

Biomass is a mixture of structural constituents (hemicelluloses, cellulose, and lignin) and minor amounts of extractives each of which pyrolyzes at different rates and by different mechanisms and pathways. All biomass materials can be converted into energy via thermochemical and biological processes. Of all the thermochemical conversion technologies, biomass gasification has attracted the most interest as it offers higher efficiencies in relation to combustion, whereas flash pyrolysis is still in the development stage (Demirbas and Arin, 2002). It is believed that as the reaction progresses the carbon becomes less reactive and forms stable chemical structures, and consequently the activation energy increases as the conversion level of biomass increases (Tran and Charanjit, 1978).

Biomass gasification can be considered as a form of pyrolysis, which takes place in higher temperatures and produces a mixture of gases with H_2 content ranging between 6 and 6.5% (McKendry, 2002; Bridgwater, 2003). Hydrogen can be produced from biomass via two thermochemical processes: (1) gasification followed by reforming of the syngas and (2) fast pyrolysis followed by reforming of the carbohydrate fraction of the bio-oil. In each process, water-gas shift is used to convert the reformed gas into hydrogen and pressure swing adsorption is used to purify the product.

Power generation from gaseous products from biomass gasification has been found to be the most promising biomass gasification technology. Gasification technologies provide the opportunity to convert renewable biomass feedstocks into clean fuel gases or synthesis gases. The synthesis gases include mainly hydrogen and carbon monoxide (H_2 + CO), which is also known as syngas. Biosyngas is a gas rich in CO and H_2 obtained by gasification of biomass (Maschio *et al.*, 1994).

Currently, hydrogen is most economically produced from natural gas. The most studied technology for hydrogen production is steam-methane reforming, in which methane reacts with steam to produce a mixture of hydrogen, carbon dioxide, and carbon monoxide.

2.4.1 Steam-reforming Process

In the steam-reforming reaction, steam reacts with hydrocarbons in the feed to produce mainly carbon monoxide and hydrogen, commonly called synthesis gas. Steam reforming can be applied to various solid waste materials including municipal organic waste, waste oil, sewage sludge, paper mill sludge, black liquor, refuse-derived fuel, and agricultural waste. Steam reforming of natural gas, sometimes referred to as steam methane reforming, is the most common method of producing commercial bulk hydrogen. Steam reforming of natural gas is currently the least expensive method of producing hydrogen and is used for about half of the world's production of hydrogen.

Hydrogen production from carbonaceous solid wastes requires multiple catalytic reaction steps: for the production of high-purity hydrogen, the reforming of fuels is followed by two water-gas-shift reaction steps, a final carbon monoxide purification, and carbon dioxide removal. Steam reforming, partial oxidation, and autothermal reforming of methane are well-developed processes for the production of hydrogen. Stepwise steam reforming of methane for the production of carbon-monoxide-free hydrogen has been investigated at various process conditions by Choudhary and Goodman (2000). The process consists of two steps involving the decomposition of methane to carbon-monoxide-free hydrogen and surface carbon in the first step followed by steam gasification of this surface carbon in the second step. The carbon-monoxide-free hydrogen formed in the first step is produced in the second step of the reaction. The mixture of gases can be separated and methane-rich gas mixture returned to the first step (Choudhary and Goodman, 2000). Steam, at high temperatures (975 to 1375 K), is mixed with methane gas in a reactor with a nickel-based catalyst at 3 to 25 bar pressure to yield carbon monoxide (CO) and hydrogen (H_2). Steam reforming is the process by which methane and other hydrocarbons in natural gas are converted into hydrogen and carbon monoxide by reaction with steam over a nickel catalyst on a ceramic support. The hydrogen and carbon monoxide are used as initial material for other industrial processes:

$$CH_4 + H_2O \leftrightarrows CO + 3H_2 \quad \Delta H = +251\ \text{kJ/mol}. \tag{2.8}$$

It is usually followed by the shift reaction:

$$CO + H_2O \leftrightarrows CO_2 + H_2 \quad \Delta H = -42\,kJ/mol. \tag{2.9}$$

The theoretical percentage of hydrogen to water is 50%. The further chemical reaction for most hydrocarbons that take place is:

$$C_nH_m + n\,H_2O \leftrightarrows nCO + (m/2 + n)\,H_2. \tag{2.10}$$

It is possible to increase the efficiency to over 85% with an economic profit at higher thermal integration. There are two types of steam reformers for small-scale hydrogen production: conventional reduced-scale reformers and specially designed reformers for fuel cells.

Commercial catalysts consist essentially of nickel supported on alumina. In one study, in the conversion of cyclohexane, magnesiuminhibited the formation of hydrogenolysis products. Nonetheless, the presence of calcium did not influence the metallic phase. The impregnated Ni/MgO catalyst performed better than the other types (Santos *et al.*, 2004).

Compared with other biomass thermochemical gasification processes such as air gasification or steam gasification, supercritical water gasification can directly deal with wet biomass without drying and have high gasification efficiency in lower temperatures. The cost of hydrogen production from supercritical water gasification of wet biomass is several times higher than the current price of hydrogen from steam methane reforming. In one study, biomass was gasified in supercritical water at a series of temperatures and pressures during different resident times to form a product gas composed of H_2, CO_2, CO, CH_4, and a small amount of C_2H_4 and C_2H_6 (Demirbas, 2004).

The yield of hydrogen from conventional pyrolysis of corncob increases from 33 to 40% when the temperature is increased from 775 to 1025 K. The yields of hydrogen from steam gasification increase from 29% to 45% for (water/solid) = 1 and from 29% to 47% for (water/solid) = 2 when the temperature is increased from 975 to 1225 K (Demirbas, 2006c). The pyrolysis is carried out at moderate temperatures and steam gasification at the highest temperatures.

2.4.2 Fuels from Bio-syngas via Fischer–Tropsch Synthesis

The Fischer–Tropsch synthesis (FTS) for the production of liquid hydrocarbons from coal-based synthesis gas has been the subject of renewed interest for conversion of coal and natural gas into liquid fuels (Jin and Datye, 2000). The use of iron-based catalysts is attractive due to their high FTS activity as well as their water-gas shift reactivity, which helps make up the deficit of H_2 in the syngas from modern energy-efficient coal gasifiers (Rao *et al.*, 1992). The FTS for the production of transportation fuels and other chemicals from synthesis has attracted much attention due to the pressure from oil supplies (Dry, 1999). The interest in iron-based catalysts stems from their relatively low cost and excellent water-gas-shift (WGS) reaction activity, which helps to make up the deficit of H_2 in the syngas from coal gasification (Wu *et al.*, 2004; Jothimurugesan *et al.*, 2000; Jun *et al.*, 2004). Hydrocarbon synthesis from biomass-derived syngas (bio-syngas) has been investigated as a potential way to use biomass. Only biomass offers the possibility to produce liquid, carbon-neutral transportation fuels (Tijmensen *et al.*, 2002). The FTS is used to produce chemicals, gasoline, and diesel fuel. The FT products are predominantly linear; hence the quality of the diesel fuel is very high. Since purified synthesis gas is used in FTS, all the products are sulfur and nitrogen free (Dry, 2002a). Given suitable economic conditions, FTS is an alternative route to liquid fuels and chemicals. Being sulfur and nitrogen free and low in aromatics, the fuels are more environmentally friendly than those produced from crude oil. In particular, the production of environmentally friendly high-quality diesel fuel is an attractive application of FTS (Dry, 1999).

Franz Fischer and Hans Tropsch first studied the conversion of syngas into larger, useful organic compounds in 1923 (Spath and Mann, 2000). Using syngas made from coal, they were able to produce liquid hydrocarbons rich in oxygenated compounds in what was termed the synthol process. Following these initial discoveries, considerable effort went into developing improved and more selective catalysts for this process. The process of converting CO and H_2 mixtures into liquid hydrocarbons over a transition-metal catalyst has become known as the Fischer–Tropsch synthesis (FTS). The first FTS plants began operation in Germany in 1938 but closed down after the Second World War. Then in 1955, Sasol, a world leader in the commercial production of liquid fuels and chemicals from coal and crude oil, started Sasol I in Sasolburg, South Africa. Following the success of Sasol I, Sasol II and Sasol III, located in Secunda, South Africa, came online in 1980 and 1982, respectively (Spath and Mann, 2000; Spath and Dayton, 2003). The FTS is an essential step in the conversion of carbon-containing feedstocks to liquid fuels such as diesel. Major advantages of the FTS are (1) its flexibility in feedstocks (natural gas, coal, biomass), (2) the large and even sustainable resources involved in the process, (3) the ultraclean (low sulfur content) products that result, and (4) its suitability for converting difficult-to-process resources. A major drawback of the FTS is the polymerizationlike nature of the process, yielding a wide product spectrum, ranging from compounds with low molecular mass like methane to products with very high molecular mass like heavy waxes.

The FTS-based gas-to-liquid (GTL) technology includes three processing steps: syngas generation, syngas conversion, and hydroprocessing. To make the GTL technology more cost effective, the focus must be on reducing both the capital and the operating costs of gas-to-liquid plants (Vosloo, 2001). For some time now the price has been up to \$60 per barrel. It has been estimated that the FT process should be viable at crude oil prices of about \$20 per barrel (Jager, 1998). The current commercial applications of the FT process are geared at the production of the valuable linear alpha olefins and of fuels such as LPG, gasoline, kerosene, and diesel. Since the FT process produces predominantly linear hydrocarbons, the production of high-quality diesel fuel is currently of considerable interest (Dry, 2004). The most expensive section of an FT complex is the production of purified syngas, and so its composition should match the overall usage ratio of the FT reactions, which in turn depends on the product selectivity (Dry, 2002a). The industrial application of the FT process started in Germany, and by 1938 there were nine plants in operation having a combined capacity of about 660×10^3 t per year (Anderson, 1984)

According to operating conditions, FTS always produces a wide range of olefins, paraffins, and oxygenated products (alcohols, aldehydes, acids, and ketones). The variables that influence the spread of the products are temperature, feed gas composition, pressure, catalyst type, and promoters (Dry, 2002b).

The high-temperature fluidized-bed FT reactors with iron catalyst are ideal for the production of large amounts of linear olefins. As petrochemicals they sell at much higher prices than fuels. The olefin content of the C_3, C_5–C_{12}, and C_{13}–C_{18} cuts are typically 85, 70, and 60%, respectively (Dry, 2002b).

The Al_2O_3/SiO_2 ratio has significant influence on iron-based catalyst activity and selectivity in the process of FTS. Product selectivities also change significantly with different Al_2O_3/SiO_2 ratios. The selectivity of low-molecular-weight hydrocarbons increases and the olefin-to-paraffin ratio in the products shows a monotonic decrease with an increasing Al_2O_3/SiO_2 ratio (Jothimurugesan *et al.*, 2000). Recently, Jun *et al.* (2004) studied FTS over Al_2O_3– and SiO_2–supported iron-based catalysts from biomass-derived syngas. They found that Al_2O_3 as a structural promoter facilitated the better dispersion of copper and potassium and gave much higher FTS activity.

More recently, there has been some interest in the use of FTS for biomass conversion into synthetic hydrocarbons. Bio-syngas consists mainly of H_2, CO, CO_2, and CH_4. Although the composition of bio-syngas is not suitable for direct use in FTS, it can be tailored by methane reforming, water-gas-shift reaction, and CO_2 removal. To maximize the utilization of carbon sources, the steam reforming of bio-syngas with additional natural gas feedstock can be considered (Bukur *et al.*, 1995; Dong and Steinberg, 1997; Specht *et al.*, 1999; Larson and Jin, 1999; Lee *et al.*, 2001). Hydrocarbon synthesis from bio-syngas has been investigated as a potential way to use biomass. FTS has been carried out using a $CO/CO_2/H_2/Ar$ (11/32/52/5 vol.%) mixture as a model for bio-syngas on coprecipitated Fe/Cu/K, Fe/Cu/Si/K, and Fe/Cu/Al/K catalysts in a fixed-bed reactor. Some performances of the catalysts that depend on the syngas composition have also been presented (Jun *et al.*, 2004).

To produce bio-syngas from a biomass fuel the following procedures are necessary: (a) gasification of the fuel, (b) cleaning the product gas, (c) using the synthesis gas to produce chemicals, and (d) using the synthesis gas as an energy carrier in fuel cells. Figure 2.4 shows how liquid fuels are obtained from coal, biomass, and natural gas by FTS.

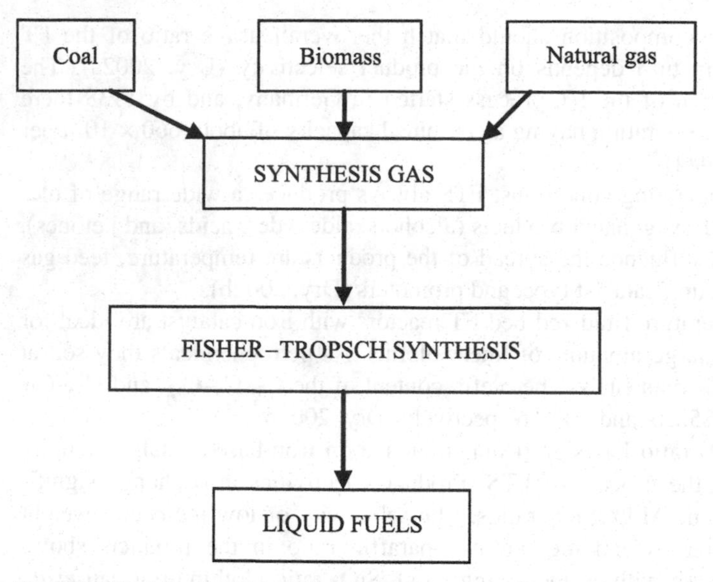

Fig. 2.4 Obtaining liquid fuels from coal, biomass, and natural gas by FTS

Figure 2.5 shows the production of diesel fuel from bio-syngas by FTS. Bio-syngas is a gas rich in CO and H_2 obtained by gasification of biomass. Biomass can be converted into bio-syngas by non-catalytic, catalytic, and steam-gasification processes.

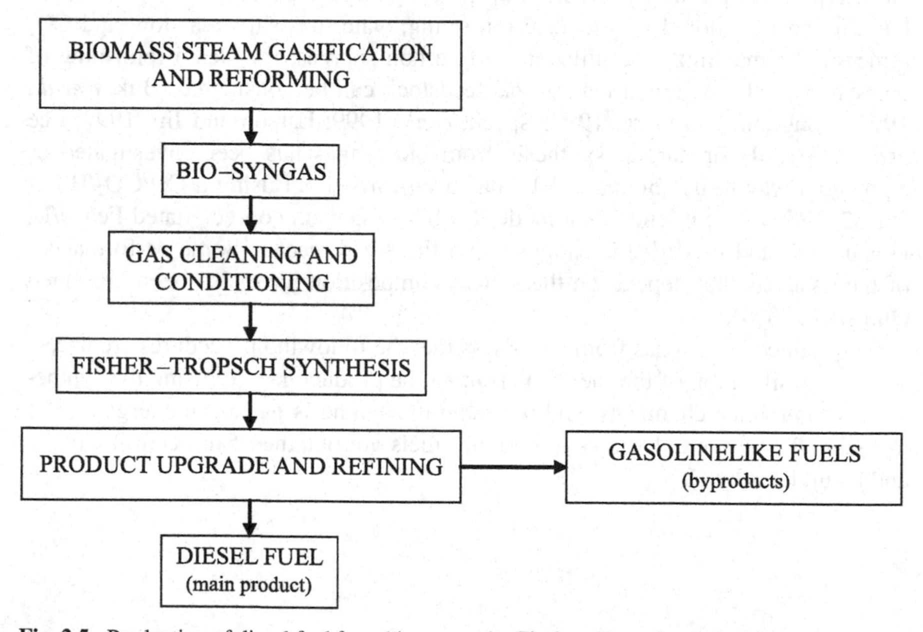

Fig. 2.5 Production of diesel fuel from bio-syngas by Fischer–Tropsch synthesis

FTS was established in 1923 by German scientists Franz Fischer and Hans Tropsch. The main aim of FTS is the synthesis of long-chain hydrocarbons from a CO and H_2 gas mixture. The FTS is described by the following set of equations (Schulz, 1999; Li *et al.*, 2002):

Basic FTS reactions:

$$nCO + 2nH_2 \rightarrow (-CH_2-) + nH_2O, \tag{2.11}$$

$$nCO + (2n+1)H_2 \rightarrow C_nH_{2n+1} + nH_2O, \tag{2.12}$$

$$nCO + (n + m/2)\ H_2 \rightarrow C_nH_m + nH_2O, \tag{2.13}$$

where *n* is the average length of the hydrocarbon chain and *m* is the number of hydrogen atoms per carbon. All reactions are exothermic, and the product is a mixture of different hydrocarbons in that paraffin and olefins are the main parts.

In FTS one mole of CO reacts with two moles of H_2 in the presence of a cobalt (Co)-based catalyst to yield a hydrocarbon chain extension ($-CH_2-$). The reaction of the synthesis is exothermic ($\Delta H = -165$ kJ/mol):

$$CO + 2H_2 \rightarrow -CH_2- + H_2O \quad \Delta H = -165\ \text{kJ/mol}. \tag{2.14}$$

The $-CH_2-$ is a building block for longer hydrocarbons. A main characteristic regarding the performance of FTS is the liquid selectivity of the process (Stelmachowski and Nowicki, 2003). For this reaction, given by Eq. (2.7), an H_2/CO ratio of at least 2 is necessary for the synthesis of the hydrocarbons. When iron (Fe)-based catalysts are used with water-gas-shift reaction activity, the water produced in reaction (2) can react with CO to form additional H_2. The reaction of the synthesis is exothermic ($\Delta H = -204$ kJ/mol). In this case a minimal H_2/CO ratio of 0.7 is required:

$$2CO + H_2 \rightarrow -CH_2- + CO_2 \quad \Delta H = -204\ \text{kJ/mol}. \tag{2.15}$$

Typical operating conditions for FTS are a temperature range of 475 to 625 K and pressures of 15 to 40 bar, depending on the process. The kind and quantity of liquid product obtained is determined by the reaction temperature, pressure, and residence time, the type of reactor, and the catalyst used. Catalysts and reactors have been extensively investigated for liquid-phase FTS (Davis, 2002). Iron catalysts have a higher tolerance for sulfur, are cheaper, and produce more olefin products and alcohols. However, the lifetime of iron catalysts is short and in commercial installations generally limited to 8 weeks.

The design of a biomass gasifier integrated with a FTS reactor must be aimed at achieving a high yield of liquid hydrocarbons. For the gasifier, it is important to avoid methane formation as much as possible and convert all carbon in the biomass to mainly carbon monoxide and carbon dioxide (Prins *et al.*, 2004). Gas cleaning is an important process before FTS. Gas cleaning is even more important for the integration of a biomass gasifier and a catalytic reactor. To avoid poisoning of FTS catalysts, tar, hydrogen sulfide, carbonyl sulfide, ammonia, hydrogen cyanide, alkalis, and dust particles must be removed thoroughly (Stelmachowski and Nowicki, 2003).

Synthetic FTS diesel fuels can have excellent autoignition characteristics. FTS diesel is composed of only straight-chain hydrocarbons and has no aromatics or sulfur. Reaction parameters are temperature, pressure, and H_2/CO ratio. FTS product composition is strongly influenced by catalyst composition: product from a cobalt catalyst is higher in paraffins and product from an iron catalyst is higher in olefins and oxygenates (Demirbas, 2006a).

2.5 Biodiesel

Vegetable oil (m)ethyl esters, commonly referred to as "biodiesel", are prominent candidates as alternative diesel fuels. The name biodiesel has been given to transesterified vegetable oil to describe its use as a diesel fuel (Demirbas, 2002). There has been renewed interest in the use of vegetable oils for making biodiesel due to its less polluting and renewable nature as opposed to conventional diesel, which is a fossil fuel that can be depleted (Ghadge and Raheman, 2006). Biodiesel is technically competitive with or offers technical advantages over conventional petroleum diesel fuel. Vegetable oils can be converted into their (m)ethyl esters via a transesterification process in the presence of a catalyst. Methyl, ethyl, 2-propyl, and butyl esters have been prepared from vegetable oils through transesterification using potassium and/or sodium alkoxides as catalysts. The purpose of the transesterification process is to lower the viscosity of the oil. Ideally, transesterification is potentially a less expensive way of transforming the large, branched molecular structure of bio-oils into smaller, straight-chain molecules of the type required in regular diesel combustion engines.

Biodiesel esters are characterized by their physical and fuel properties including density, viscosity, iodine value, acid value, cloud point, pure point, gross heat of combustion, and volatility. Biodiesel fuels produce slightly lower power and torque and consume more fuel than No. 2 diesel (D2) fuel. Biodiesel is better than diesel fuel in terms of sulfur content, flash point, aromatic content, and biodegradability (Bala, 2005).

The cost of biodiesels varies depending on the base stock, geographic area, variability in crop production from season to season, the price of crude petroleum, and other factors. Biodiesel is more than twice as expensive as petroleum diesel. The high price of biodiesel is in large part due to the high price of the feedstock. However, biodiesel can be made from other feedstocks, including beef tallow, pork lard, and yellow grease (Demirbas, 2005).

Most of the biodiesel currently made uses soybean oil, methanol, and an alkaline catalyst. The high value of soybean oil as a food product makes production of a cost-effective fuel very challenging. However, there are large amounts of low-cost oils and fats such as restaurant waste and animal fats that could be converted into biodiesel. The problem with processing these low-cost oils and fats is that they often contain large amounts of free fatty acids (FFA) that cannot be converted into biodiesel using an alkaline catalyst (Demirbas, 2003; Canakci and Van Gerpen, 2001).

Biodiesel is an environmentally friendly alternative liquid fuel that can be used in any diesel engine without modification. There has been renewed interest in the use of vegetable oils for making biodiesel due to its less polluting and renewable nature compared with conventional petroleum diesel fuel. If the biodiesel valorized efficiently at energy purpose, so would be benefit for the environment and the local population, job creation, provision of modern energy carriers to rural communities.

2.6 Bio-oil

The term bio-oil is used mainly to refer to liquid fuels. There are several reasons why bio-oils are considered relevant technologies by both developing and industrialized countries. They include energy security reasons, environmental concerns, foreign exchange savings, and socioeconomic issues related to the rural sector.

Bio-oils are liquid or gaseous fuels made from biomass materials such as agricultural crops, municipal wastes, and agricultural and forestry byproducts via biochemical or thermochemical processes. They can replace conventional fuels in vehicle engines, either totally or partially in a blend (EC, 2004). The organic fraction of almost any form of biomass, including sewage sludge, animal waste, and industrial effluents, can be broken down through anaerobic digestion into a methane and carbon dioxide mixture called "biogas". Biogas is an environment friendly, clean, cheap, and versatile fuel (Kapdi *et al.*, 2005).

Pyrolysis/cracking is defined as the cleavage to smaller molecules by thermal energy. Hydrogen can be produced economically from woody biomass (Encinar *et al.*, 1998). Biomass can be thermally processed through gasification or pyrolysis to produce hydrogen. The main gaseous products from biomass are the following (Wang *et al.*, 1997):

$$\text{Pyrolysis of biomass} \rightarrow H_2 + CO_2 + CO + \text{Gaseous and liquid hydrocarbons}, \quad (2.16)$$

$$\text{Catalytic steam reforming of biomass} \rightarrow H_2 + CO_2 + CO, \quad (2.17)$$

$$\text{FT synthesis of } (H_2 + CO) \rightarrow \text{Gaseous and liquid hydrocarbons}. \quad (2.18)$$

If the purpose were to maximize the yield of liquid products resulting from biomass pyrolysis, a low temperature, high heating rate, short gas residence time process would be required. For a high char production, a low temperature, low heating rate process would be chosen (Bridgwater, 2003). If the purpose were to maximize the yield of fuel gas resulting from pyrolysis, a high temperature, low heating rate, long gas residence time process would be preferred (Encinar *et al.*, 1998).

Fischer–Tropsch liquids from natural gas and ethanol from biomass may become widespread. The Fischer–Tropsch liquids will compete with petroleum if natural gas is imported at very low prices and the sulfur content of the liquids is much lower than that of petroleum diesel fuel (MacLean and Lave, 2003). Figure 2.6 shows the production facilities of green diesel and green gasoline and other fuels from bio-syngas by FTS.

FTS can be carried out in a supercritical fluid medium (SFM). When hexane is used as the fluid, with increasing pressure in the supercritical medium, the density and heat capacity of the hexane-dominated phase increase. The decrease in mass transfer rates at the higher pressure is offset somewhat by the increase in the intrinsic reaction rates. At a space velocity of 135 g hexane/g catalyst/h, end of run (8 h) isomerizations are roughly twofold higher and deactivation rates are threefold lower in near-critical reaction mixtures when compared to subcritical reaction mixtures (Balat, 2006).

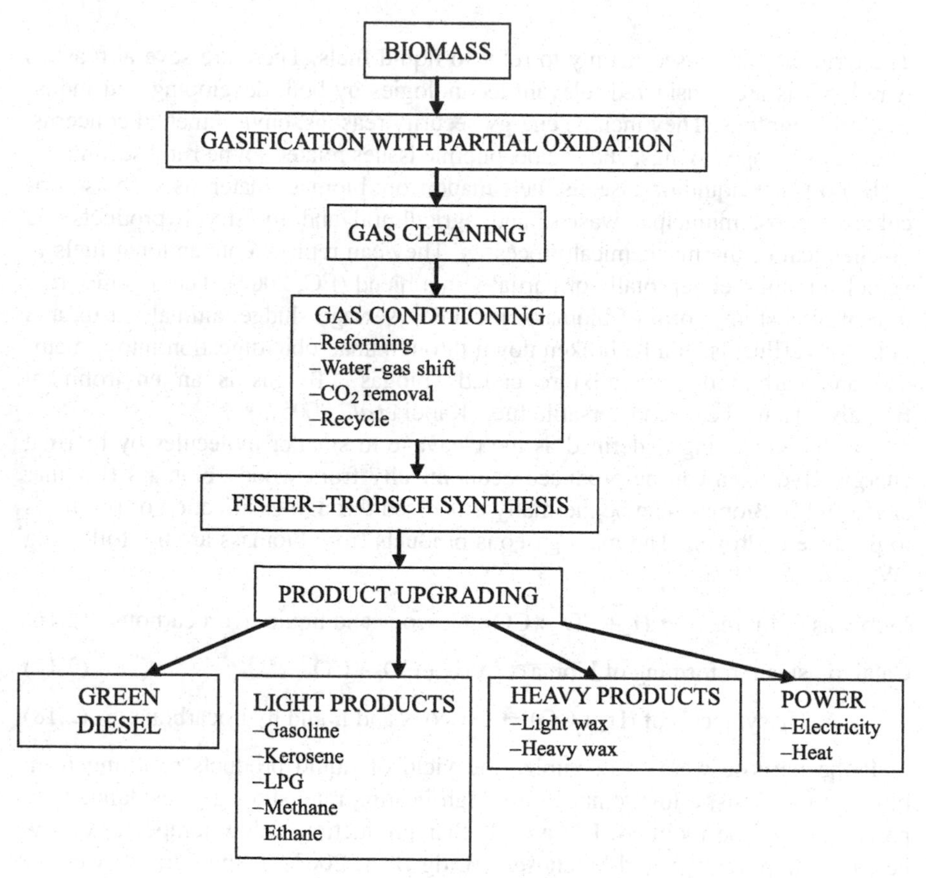

Fig. 2.6 Green diesel and green gasoline facilities from biomass via Fischer–Tropsch synthesis

2.7 Global Biofuel Scenarios

Renewable resources are more evenly distributed than fossil and nuclear resources, and energy flows from renewable resources are more than three orders of magnitude higher than current global energy use. Today's energy system is unsustainable because of uncompeting issues as well as environmental, economic, and geopolitical concerns that have implications far into the future (UNDP, 2000).

According to the International Energy Agency (IEA), scenarios developed for the USA and the EU indicate that near-term targets of up to 6% displacement of petroleum fuels with biofuels appear feasible using conventional biofuels, given available cropland. A 5% displacement of gasoline in the EU would require about 5% of available cropland to produce ethanol, while in the USA 8% would be required. A 5% displacement of diesel would require 13% of US cropland and 15% in the EU. The recent commitment by the US government to increase bioenergy threefold in 10 years has added impetus to the search for viable biofuels (IEA, 2004).

Dwindling fossil fuel stocks and the increasing dependency of the USA on imported crude oil have led to a major interest in expanding the use of bioenergy. The EU has also adopted a proposal for a directive promoting the use of biofuels with measures ensuring that biofuels would account for at least 2% of the market for gasoline and diesel sold as transport fuel by the end of 2005, increasing in stages to a minimum of 5.75% by the end of 2010 (Hansen *et al.*, 2005). Biomass can be converted into biofuels such as bioethanol and biodiesel and thermochemical conversion products such as syn-oil, bio-syngas, and biochemicals. Bioethanol is a fuel derived from renewable sources of feedstock, typically plants such as wheat, sugar beet, corn, straw, and wood. Bioethanol is a petroleum additive/substitute, and biodiesel is better than diesel fuel in terms of sulfur content, flash point, aromatic content, and biodegradability (Bala, 2005).

Figure 2.7 shows the alternative fuel projections for total automotive fuel consumption in the world. Hydrogen is currently more expensive than conventional energy sources. There are different technologies presently being applied to produce hydrogen economically from biomass. Biohydrogen technology will play a major role in the future because it can utilize renewable sources of energy (Nath and Das, 2003).

Hydrogen for fleet vehicles will probably dominate in the transportation sector in the future. To produce hydrogen via electrolysis and the transportation of liquefied hydrogen to rural areas with pipelines would be expensive. The production technology would be site specific and include steam reforming of methane and electrolysis in hydropower-rich countries. In the long run, when hydrogen is a very common energy carrier, distribution via pipeline is probably the preferred option. The cost of hydrogen distribution and refueling is very site specific.

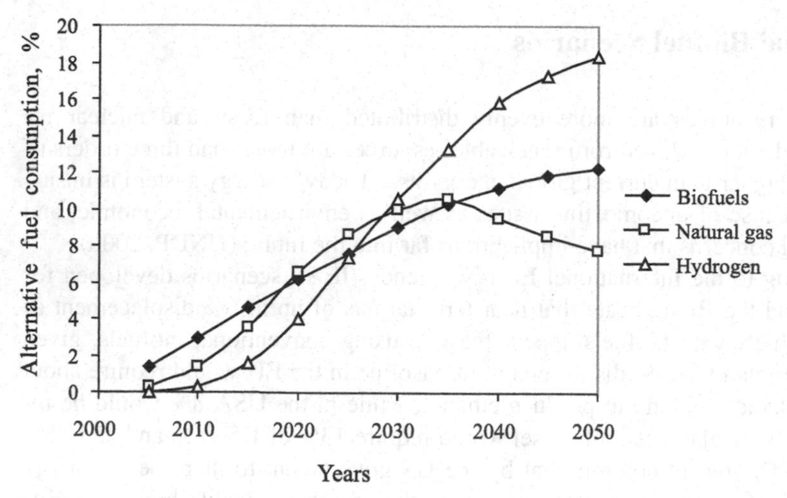

Fig. 2.7 Alternative fuel projections for total automotive fuel consumption in the world
Source: Demirbas, 2006a

References

Adeeb Z. 2004. Glycerol delignification of poplar wood chips in aqueous medium. Energy Edu Sci Technol 13:81–88

Ahón, V.R., Costa, E.F., Jr., Monteagudo, J.E.P., Fontes, C.E., Biscaia, E.C., Jr., Lage, P.L.C. 2005. A comprehensive mathematical model for the Fischer–Tropsch synthesis in well-mixed slurry reactors. Chem Eng Sci 60:677–694

Anderson, R.B. 1984. The Fischer–Tropsch Synthesis. Academic, New York

Azar, C., Lindgren, K., Andersson, B.A. 2003. Global energy scenarios meeting stringent CO_2 constraints—cost-effective fuel choices in the transportation sector. Energy Policy 31: 961–976

Bala, B.K. 2005. Studies on biodiesels from transformation of vegetable oils for diesel engines. Edu Sci Technol 15:1–45

Balat, M. 2005. Biodiesel from vegetable oils via transesterification in supercritical ethanol. Energy Edu Sci Technol 16:45–52

Balat, M. 2006. Sustainable transportation fuels from biomass materials. Energy Edu Sci Technol 17:83–103

Baltz, R.A., Burcham, A.F., Sitton, O.C., Book, N.L. 1982. The recycling of sulfuric acid and xylose in the prehydrolysis of corn stover. Energy 7:259–265

Berg, C. 1988. Towards a world ethanol market? F.O. Licht Commodity Analysis, Ratzeburg, Germany.

Bothast, R.J., Schlicher, M.A. 2005. Biotechnological processes for conversion of corn into ethanol. Appl Microbiol Biotechnol 67:19–25

Bowen, D.A., Lau, F., Zabransky, R., Remick, R., Slimane, R., Doong, S. 2003. Techno-economic analysis of hydrogen production by gasification of biomass. NREL 2003 progress report, Renewable Energy Laboratory, Golden, CO.

Bridgwater, A.V. 2003. Renewable fuels and chemicals by thermal processing of biomass. Chem Eng J 91:87–102

Brown, H.P., Panshin, A. J., Forsaith, C.C. 1952. Textbook of wood technology, Vol. II. Hill, New York

Bukur, D.B., Nowicki, L., Manne, R.V., Lang, X. 1995. Activation studies with a precipitated iron catalysts for the Fischer–Tropsch synthesis. J Catal 155:366–375

Cadenas, A., Cabezudo, S. 1998. Biofuels as sustainable technologies: perspectives for less developed countries. Technol Forecast Social Change 58:83–103

Canakci, M., Van Gerpen, J. 2001. Biodiesel production from oils and fats with high free fatty acids. Am Soc Agric Eng 4:1429–1436

Castro, F.B., Hotten, P.M., Ørskov, E.R. 1993. The potential of dilute-acid hydrolysis as a treatment for improving the nutritional quality of industrial lignocellulosic by products. Animal Feed Sci Technol 42:39–53

Chmielniak, T., Sciazko, M. 2003. Co-gasification of biomass and coal for methanol synthesis. Energy 74:393–403

Choudhary, T.V., Goodman, D.W. 2000. CO-free production of hydrogen via stepwise steam reforming of methane. J Catal 192:316–321

Davis, B.H. 2002. Overview of reactors for liquid phase Fischer–Tropsch synthesis. Catal Today 71:249–300

Demirbas, A. 1999. Fuel properties of charcoal derived from hazelnut shell and the production of briquets using pyrolytic oil. Energy 24:141–150

Demirbas, A. 2000. Mechanisms of liquefaction and pyrolysis reactions of biomass. Energy Convers Mgmt 41:633–646

Demirbas, A. 2002. Biodiesel from vegetable oils via transesterification in supercritical methanol. Energy Convers Mgmt 43:2349–2356

Demirbas, A. 2003. Biodiesel fuels from vegetable oils via catalytic and non-catalytic supercritical alcohol transesterifications and other methods: a survey. Energy Convers Mgmt 44:2093–2109

Demirbas, A. 2004. Hydrogen rich gas from fruit shells via supercritical water extraction. Int J Hydrogen Energy 29:1237–1243

Demirbas, A. 2005. Biodiesel production from vegetable oils via catalytic and non-catalytic supercritical methanol transesterification methods. Prog Energy Combust Sci 31:466–487

Demirbas, A. 2006a. Global biofuel strategies. Energy Edu Sci Technol 17:27–63.

Demirbas, A. 2006b. Biogas potential of manure and straw mixtures. Energy Sour 28:71–78

Demirbas, M.F. 2006c. Hydrogen from various biomass species via pyrolysis and steam gasification processes. Energy Sour A 28:245–252

Demirbas, A., Arin, G. 2002. An overview of biomass pyrolysis. Energy Sour 5:471–482

Demirbas, M.F., Balat, M. 2006. Recent advances on the production and utilization trends of biofuels: a global perspective. Energy Convers Mgmt 47:2371–2381

Demirbas, A., Gullu, D. 1998. Acetic acid, methanol and acetone from lignocellulosics by pyrpolysis. Edu Sci Technol 1:111–115

Difiglio, C. 1997. Using advanced technologies to reduce motor vehicle greenhouse gas emissions. Energy Policy 25:1173–1178

Dong, Y., Steinberg, M. 1997. Hynol—an economical process for methanol production from biomass and natural gas with reduced CO_2 emission. Int J Hydrogen Energy 22:971–977

Dry, M.E. 1999. Fischer–Tropsch reactions and the environment. Appl Catal A General 189:185–190

Dry, M.E. 2002a. High quality diesel via the Fischer–Tropsch process-a review. J Chem Technol Biotechnol 77:43–50

Dry, M.E. 2002b. The Fischer–Tropsch process: 1950–2000. Catal Today 71:227–241.

Dry, M.E. 2004. Present and future applications of the Fischer–Tropsch process. Appl Catal A 276:1–3.

EC (European Commission). 2004. Promoting Biofuels in Europe. European Commission, Directorate -General for Energy and Transport, B-1049 Brussels, Belgium. http://europa.eu.int/comm/dgs/energy_transport/index_en.html.

Encinar, J.M., Beltran, F.J., Ramiro, A., Gonzalez, J.F. 1998. Pyrolysis/gasification of agricultural residues by carbon dioxide in the presence of different additives: influence of variables. Fuel Process Technol 55:219–233.

Ghadge, S.V., Raheman, H. 2006. Process optimization for biodiesel production from mahua (*Madhuca indica*) oil using response surface methodology. Bioresour Technol 97:379–384.
Grassi, G. 1999. Modern bioenergy in the European Union. Renew Energy 6:985–990.
Gullu, D., Demirbas, A. 2001. Biomass to methanol via pyrolysis process. Energy Convers Mgmt 42:1349–1356.
Haas, M.J., McAloon, A.J., Yee, W.C., Foglia, T.A. 2006. A process model to estimate biodiesel production costs. Bioresour Technol 97:671–678.
Hall, D.O., Rosillo-Calle, F., Williams, R.H., Woods, J. 1993. Biomass for energy: supply prospects: Johansson, T.B., Kelly, H., Reddy, A.K.N., Williams, R.H. (eds.) Renewable Energy–for Fuels and Electricity. Island Press, Washington, D.C.
Hansen, A.C., Zhang, Q., Lyne, P.W.L. 2005. Ethanol–diesel fuel blends—a review. Technology 96:277–285.
Hao, X.H., Guo, L.J. 2002. A review on investigation of hydrogen production by biomass catalytic gasification in supercritical water. Huagong Xuebao 53:221–8 [in Chinese].
IEA (International Energy Agency). 2004. Biofuels for transport: an international perspective. 9, rue de la Fédération, 75739 Paris, cedex 15, France. www.iea.org.
IPCC. 1997. Greenhouse Gas Inventory Reference Manual: Revised 1996 IPCC Guidelines for National Greenhouse Gas Inventories, Report Vol. 3, p. 1.53, Intergovernmental Panel on Climate Change (IPCC), Paris, France. www.ipcc.ch/pub/guide.htm.
Iwasa, N., Kudo, S., Takahashi, H., Masuda, S., Takezawa, N. 1993. Highly selective supported Pd catalysts for steam reforming of methanol. Catal Lett 19:211–216.
Jager, B. 1998. Proceedings of the 5th Natural Gas Conversion Symposium, Taormina, Italy, September 1998.
Jin, Y., Datye, A.K. 2000. Phase transformations in iron Fischer–Tropsch catalysts during temperature-programmed reduction. J Catal 196:8–17.
Jothimurugesan, K., Goodwin, J.G., Santosh, S.K., Spivey, J.J. 2000. Development of Fe Fischer–Tropsch catalysts for slurry bubble column reactors. Catal Today 58:335–344.
Jun, K.W., Roh, H.S., Kim, K.S., Ryu, J.S., Lee, K.W. 2004. Catalytic investigation for Fischer–Tropsch synthesis from bio-mass derived syngas. Appl Catal A 259:221–226.
Kapdi, S.S., Vijay, V.K., Rajesh, S.K., Prasad, R. 2005. Biogas scrubbing, compression and storage: perspective and prospectus in Indian context. Renew Energy 30:1195–1202.
Kartha, S., Larson, E.D. 2000. Bioenergy primer: modernised biomass energy for sustainable development, Technical Report UN Sales Number E.00.III.B.6, United Nations Development Programme, 1 United Nations Plaza, New York, NY 10017.
Kim, S., Dale, B.E. 2005. Life cycle assessment of various cropping systems utilized for producing: bioethanol and biodiesel. Biomass Bioenergy 29:426–439.
Kumar, A., Cameron, J.B., Flynn, P.C. 2005. Pipeline transport and simultaneous saccharification of corn stover. Bioresour Technol 96:819–829.
Larson, E. D, Jin H. 1999. In: Overend, R., Chornet, E. (eds.) Proceedings of the 4th Biomass Conference of the Americas, Kidlington, UK, 29 August 1999, Elsevier, Amsterdam.
Lee, K.-W., Kim, S.-B., Jun, K.-W., Choi, M.-J. 2001. In: Williams, D.J., Durie, R.A., McMullan, P., Paulson, A.J., Smith, A.Y. (eds.) Proceedings of the 5th International Conference on Greenhouse Gas Control Technology, Cairns, Australia, 13 August 2000, CSIRO.
Li, S., Krishnamoorthy, S., Li, A., Meitzner, G.D., Iglesia, E. 2002. Promoted iron-based catalysts for the Fischer–Tropsch synthesis: design, synthesis, site densities, and catalytic properties. J Catal 206:202–217.
MacLean, H.L., Lave, L.B. 2003. Evaluating automobile fuel/propulsion system technologies. Energy Combust Sci 29:1–69.
Madras, G., Kolluru, C., Kumar, R. 2004. Synthesis of biodiesel in supercritical fluids. Fuel 83:2029–2033.
Maschio, G., Lucchesi, A., Stoppato, G. 1994. Production of syngas from biomass. Bioresour Technol 48:119–126.
Matsumura, Y., Minowa, T. 2004. Fundamental design of a continuous biomass gasification process using a supercritical water fluidized bed. Int J Hydrogen Energy 29:701–707.

May, M. 2003. Development and demonstration of Fischer–Tropsch fueled heavy-duty vehicles with control technologies for reduced diesel exhaust emissions. 9th Diesel Engine Emissions Reduction Conference. Newport, RI, 24–28 August 2003.

McKendry, P. 2002. Energy production from biomass (part 1): overview of biomass. Bioresour Technol 83:37–46.

Meher, L.C., Sagar, D.V., Naik, S.N. 2006. Technical aspects of biodiesel production by transesterification—a review. Renew Sustain Energy Rev 10:248–268.

Mirzaei, A.A., Habibpour, R., Faizi, M., Kashi, E. 2006. Characterization of iron-cobalt oxide catalysts: effect of different supports and promoters upon the structure and morphology of precursors and catalysts. Appl Catal A General 301:272–283.

Nath, K., Das, D. 2003. Hydrogen from biomass. Curr Sci 85:265–271.

Ouellette, N., Rogner, H.-H., Scott, D.S. 1997. Hydrogen-based industry from remote excess hydroelectricity. Int J Hydrogm Energy 22:397–403.

Phillips, V.D., Kinoshita, C.M., Neill, D.R., Takashi, P.K. 1990. Thermochemical production of methanol from biomass in Hawaii. Appl Energy 35:167–175.

Prakash, C.B. 1998. A critical review of biodiesel as a transportation fuel in Canada. A Technical Report. GCSI – Global Change Strategies International, Canada.

Prins, M.J., Ptasinski, K.J., Janssen, F.J.J.G. 2004. Exergetic optimisation of a production process of Fischer–Tropsch fuels from biomass. Fuel Process Technol 86:375–389.

Puhan, S., Vedaraman, N., Rambrahaman, B.V., Nagarajan, G. 2005. Mahua (*Madhuca indica*) seed oil: a source of renewable energy in India. J Sci Ind Res 64:890–896.

Puppan, D. 2002. Environmental evaluation of biofuels. Periodica Polytechnica Ser Soc Man Sci 10:95–116.

Reijnders, L. 2006. Conditions for the sustainability of biomass based fuel use. Energy Policy 34:863–876.

Rao, V.U.S., Stiegel, G.J., Cinquergrane, G.J., Srivastava, R.D. 1992. Iron-based catalysts for slurry-phase Fischer-Tropsch process: Technology review. Fuel Process Technol 30: 83–107.

Rapagna, S., Jand, N., Foscolo, P.U. 1998. Catalytic gasification of biomass to produce hydrogen rich gas. Int J Hydrogen Energy 23:551–557.

Rowell, R.M., Hokanson, A.E. 1979. Methanol from wood: a critical assessment. In: K.V. Sarkanen,. A. Tillman (eds.) Progress in Biomass Conversion. Vol. 1. Academic, New York.

Santos, D.C.R.M., Lisboa, J.S., Passos, F.B., Noronha, F.B. 2004. Characterization of steam-reforming catalysts. Braz J Chem Eng 21:203–209.

Schulz, H. 1999. Short history and present trends of FT synthesis. Appl Catal A General 186: 1–16.

Sheehan, J., Cambreco, V., Duffield, J., Garboski, M., Shapouri, H. 1998. An overview of biodiesel and petroleum diesel life cycles. A report by US Department of Agriculture and Energy, Washington, D.C., pp. 1–35.

Sie, S. T., Krishna, R. 1999. Fundamentals and selection of advanced FT-reactors. Appl Catal General 186:55–70.

Sokhansanj, S., Turhollow, A., Cushman, J., Cundiff, J. 2002. Engineering aspects of collecting corn stover for bioenergy. Biomass Bioenergy 23:347–355.

Sorensen, H.A. 1983. Energy conversion systems. Wiley, New York.

Spath, P.L., Mann, M.K. 2000. Life Cycle Assessment of hydrogen production via natural gas steam reforming. National Renewable Energy Laboratory, Golden, CO, TP-570-27637, November 2000.

Spath, P.L., Dayton, D.C. 2003. Preliminary Screening — Technical and Economic Assessment of Synthesis Gas to Fuels and Chemicals with Emphasis on the Potential for Biomass-Derived Syngas. NREL/TP-510-34929.

Specht, M., Bandi, A., Baumgart, F., Murray, C.N., Gretz, J. (eds.) 1999. Proceedings of the 4th International Conference on Greenhouse Gas Control Technologies.

Stelmachowski, M., Nowicki, L. 2003. Fuel from the synthesis gas—the role of process engineering. Energy 74:85–93.

Taherzadeh, M.J. 1999. Ethanol from Lignocellulose: Physiological effects of inhibitors and fermentation strategies. PhD thesis, Department of Chemical Reaction Engineering, Chalmers University of Technology, Göteborg, Sweden.

Takezawa, N., Shimokawabe, M., Hiramatsu, H., Sugiura, H., Asakawa, T., Kobayashi, H. 1987. reforming of methanol over Cu/ZrO_2. Role of ZrO_2 support. React Kinet Catal Lett 33:191–196.

Theander, O. 1985. In: Overand, R.P., Mile, T.A., Mudge, L.K. (eds.) Fundamentals of thermochemical biomass conversion. Elsevier, New York.

Tijmensen, M.J.A., Faaij, A.P.C., Hamelinck, C.N., van Hardeveld, R.M.R. 2002. Exploration of the possibilities for production of Fischer Tropsch liquids and power via biomass gasification. Biomass Bioenergy 23:129–152.

Tran, D.Q., Charanjit, R. 1978. A kinetic model for pyrolysis of Douglas fir bark. Fuel 57:293–298

UNDP (United Nations Development Programme). 2000. World Energy Assessment. Energy and the Challenge of Sustainability, New York.

Usta, N., Ozturk, E., Can, O., Conkur, E.S., Nas, S., Con, A.H., Can, A.C., Topcu, M. 2005. of biodiesel fuel produced from hazelnut soapstock/waste sunflower oil mixture in a diesel engine. Energy Convers Mgmt 46:741–755.

Vasudevan, P., Sharma, S., Kumar, A. 2005. Liquid fuel from biomass: an overview. J Sci Ind Res 64:822–831.

Vosloo, A.C. 2001. Fischer–Tropsch: a futuristic view. Fuel Process Technol 71:149–155.

Wang, D., Czernik, S., Chornet, E. 1998. Production of hydrogen from biomass by catalytic steam reforming of fast pyrolysis oils. Energy Fuels 12:19–24.

Wang, D., Czernik, S., Montane, D., Mann, M., Chornet, E. 1997. Biomass to hydrogen via fast pyrolysis and catalytic steam reforming of the pyrolysis oil or its fractions. Ind Eng Chem Res 36:1507–1518.

Wu, B.S., Bai, L., Xiang, H.W., Li, Y.W., Zhang, Z.X., Zhong, B. 2004. An active iron catalyst containing sulfur for Fischer–Tropsch synthesis. Fuel 83:205–512.

Yoshida, T., Oshima, Y., Matsumura, Y. 2004. Gasification of biomass model compounds and real biomass in supercritical water. Biomass Bioenergy 26:71–78.

Chapter 3
Vegetable Oils and Animal Fats

3.1 Use of Vegetable Oils and Animal Fats in Fuel Engines

Vegetable oils have become more attractive recently because of their environmental benefits and the fact that they are made from renewable resources. More than 100 years ago, Rudolph Diesel tested vegetable oil as the fuel for his engine (Shay, 1993). Vegetable oils have the potential to replace a fraction of the petroleum distillates and petroleum-based petrochemicals in the near future. Vegetable-oil fuels presently do not compete with petroleum-based fuels because they are more expensive. However, with recent increases in petroleum prices and uncertainties surrounding petroleum availability, there is renewed interest in using vegetable oils in diesel engines. The diesel boiling range is of particular interest because it has been shown to reduce particulate emissions significantly relative to petroleum diesel fuel (Giannelos *et al.*, 2002).

Chemically speaking, vegetable oils and animal fats are triglyceride molecules in which three fatty acid groups are esters attached to one glycerol molecule (Gunstone and Hamilton, 2001). Fats and oils are primarily water-insoluble, hydrophobic substances in the plant and animal kingdoms that are made up of one mole of glycerol and three moles of fatty acids and are commonly referred to as triglycerides (Sonntag, 1979).

Chemically, fats and oils are carboxylic esters derived from the single alcohol glycerine and are known as triglycerides. Triglycerides derive from many different carboxylic acids. Triglyceride molecules differ in the nature of the alkyl chain bound to glycerol. The proportions of the various acids vary from fat to fat; each fat has its characteristic composition. Although thought of as esters of glycerine and a varying blend of fatty acids, in fact these oils contain free fatty acids and diglycerides as well. Triglyceride vegetable oils and fats include not only edible but also inedible vegetable oils and fats such as linseed oil, castor oil, and tung oil, used in lubricants, paints, cosmetics, pharmaceuticals, and other industrial purposes.

More than 350 oil-bearing crops have been identified, of which only soybean, palm, sunflower, safflower, cottonseed, rapeseed, and peanut oils are considered

Table 3.1 Oil species for biofuel production

Group	Source of oil
Major oils	Coconut (Copra), corn (maize), cottonseed, canola (a variety of rapeseed), olive, peanut (groundnut), safflower, sesame, soybean, and sunflower
Nut oils	Almond, cashew, hazelnut, macadamia, pecan, pistachio and walnut
Other edible oils	Amaranth, apricot, argan, artichoke, avocado, babassu, bay laurel, beech nut, ben, Borneo tallow nut, carob pod (algaroba), cohune, coriander seed, false flax, grape seed, hemp, kapok seed, lallemantia, lemon seed, macauba fruit (*Acrocomia sclerocarpa*), meadowfoam seed, mustard, okra seed (hibiscus seed), perilla seed, pequi, (*Caryocar brasiliensis* seed), pine nut, poppyseed, prune kernel, quinoa, ramtil (*Guizotia abyssinica* seed or Niger pea), rice bran, tallow, tea (camellia), thistle (*Silybum marianum* seed), and wheat germ
Inedible oils	Algae, babassu tree, copaiba, honge, jatropha or ratanjyote, jojoba, karanja or honge, mahua, milk bush, nagchampa, neem, petroleum nut, rubber seed tree, silk cotton tree, and tall
Other oils	Castor, radish, and tung

Table 3.2 World vegetable and marine oil consumption (million metric ton)

Oil	1998	1999	2000	2001	2002	2003
Soybean	23.5	24.5	26.0	26.6	27.2	27.9
Palm	18.5	21.2	23.5	24.8	26.3	27.8
Rapeseed	12.5	13.3	13.1	12.8	12.5	12.1
Sunflower seed	9.2	9.5	8.6	8.4	8.2	8.0
Peanut	4.5	4.3	4.2	4.7	5.3	5.8
Cottonseed	3.7	3.7	3.6	4.0	4.4	4.9
Coconut	3.2	3.2	3.3	3.5	3.7	3.9
Palm kernel	2.3	2.6	2.7	3.1	3.5	3.7
Olive	2.2	2.4	2.5	2.6	2.7	2.8
Fish	1.2	1.2	1.2	1.3	1.3	1.4
Total	80.7	85.7	88.4	91.8	95.1	98.3

potential alternative fuels for diesel engines (Goering *et al.*, 1982; Pryor *et al.*, 1982). Table 3.1 shows the oil species that can be used in biodiesel production. Worldwide consumption of soybean oil was the highest in 2003 (27.9 million metric tons). Table 3.2 shows the world vegetable and marine oil consumption between 1998 and 2003.

Vegetable oils are a renewable and potentially inexhaustible source of energy with energy content close to diesel fuel. On the other hand, extensive use of vegetable oils may cause other significant problems such as starvation in developing countries.

Vegetable-oil fuels have not been acceptable because they are more expensive than petroleum fuels. However, with recent increases in petroleum prices and uncertainties surrounding petroleum availability, vegetable oils have become more attractive recently because of their environmental benefits and the fact that they are made from renewable resources (Demirbas, 2003b; Giannelos *et al.*, 2002).

The first use of vegetable oils as a fuel was in 1900. The advantages of vegetable oils as diesel fuel are liquidity, ready availability, renewability, lower sulfur and aromatic content, and biodegradability (Goering *et al.*, 1982). The main disadvantages of vegetable oils as diesel fuel are higher viscosity, lower volatility, and the reactivity of unsaturated hydrocarbon chains. The problems met in long-term engine tests, according to results obtained by earlier researchers (Komers *et al.*, 2001; Darnoko and Cheryan, 2000), may be classified as follows: coking on injectors, more carbon deposits, oil ring sticking, and thickening and gelling of the engine lubricant oil. All vegetable oils are extremely viscous, with viscosities ranging from 10 to 17 times greater than D2 fuel (D2 fuel is a diesel engine fuel with 10 to 20 carbon number hydrocarbons) (Srivastava and Prasad, 2000).

A variety of biolipids can be used to produce biodiesel. These are (a) virgin vegetable oil feedstock; rapeseed and soybean oils are most commonly used, though other crops such as mustard, palm oil, sunflower, hemp, and even algae show promise; (b) waste vegetable oil; (c) animal fats including tallow, lard, and yellow grease; and (d) non-edible oils such as jatropha oil, neem oil, mahua oil, castor oil, tall oil, *etc*.

Soybeans are commonly used in the United States for food products, which has led to soybean biodiesel's becoming the primary source for biodiesel in that country. In Malaysia and Indonesia palm oil is used as a significant biodiesel source. In Europe, rapeseed is the most common base oil used in biodiesel production. In India and southeast Asia, the jatropha tree is used as a significant fuel source.

Algae can grow practically anywhere there is enough sunlight. Some algae can grow in saline water. The most significant difference of algal oil is in the yield and, hence, its biodiesel yield. According to some estimates, the yield (per acre) of oil from algae is over 200 times the yield from the best-performing plant/vegetable oils (Sheehan *et al.*, 1998). Microalgae are the fastest growing photosynthesizing organisms. They can complete an entire growing cycle every few days. Approximately 46 tons of oil/hectare/year can be produced from diatom algae. Different algae species produce different amounts of oil. Some algae produce up to 50% oil by weight. The production of algae to harvest oil for biodiesel has not been undertaken on a commercial scale, but working feasibility studies have been conducted to arrive at the above number.

Specially bred mustard varieties can produce reasonably high oil yields and have the added benefit that the meal left over after the oil has been pressed out can act as an effective and biodegradable pesticide.

3.2 Vegetable Oil Resources

World annual petroleum consumption and vegetable oil production is about 4.018 and 0.107 billion tons, respectively. Global vegetable oil production increased from 56 million tons in 1990 to 88 million tons in 2000, following a below-normal increase. World vegetable and marine oil consumption is tabulated in Table 3.2. Figure 3.1 shows the plots of percentages of world oil consumption by year. Figure 3.2 shows the total global production and consumption of vegetable oil by year. Leading the gains in vegetable oil production was a recovery in world palm oil output, from 18.5 million tons in 1998 to 27.8 million in 2003.

The major exporters of vegetable oils are Malaysia, Argentina, Indonesia, the Philippines, and Brazil. The major importers of vegetable oils are China, Pakistan, Italy, and the United Kingdom. A few countries such as the Netherlands, Germany, the United States, and Singapore are both major exporters as well as importers of vegetable oils (Bala, 2005).

Global vegetable oil consumption rose modestly from 79.5 million tons in 1998 to 96.9 million in 2003. A large portion of the gain occurred in India, where even small price shifts can cause a substantial change in consumption. Indian palm oil

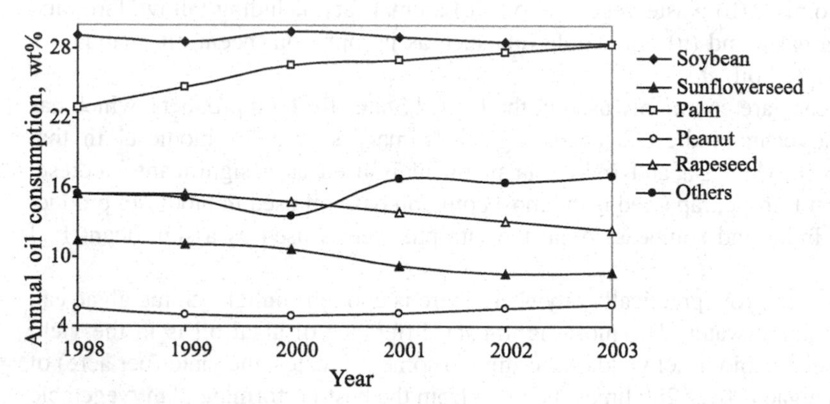

Fig. 3.1 Percentage of world oil consumption by year

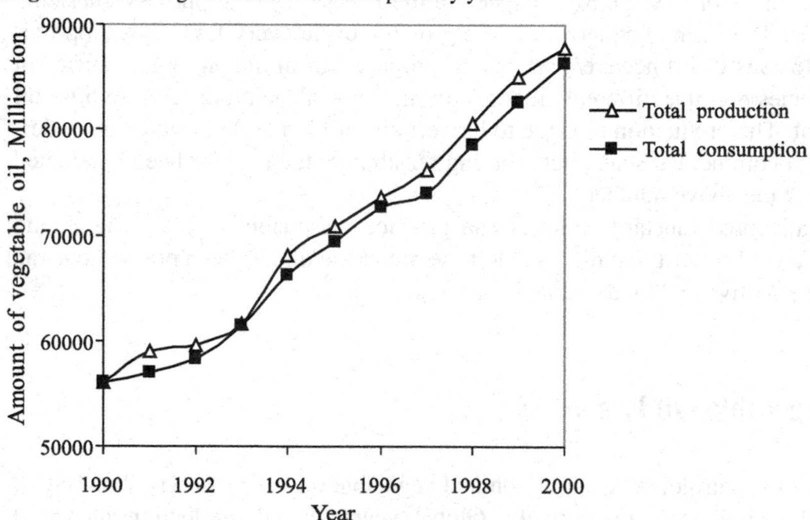

Fig. 3.2 Total global production and consumption of vegetable oil by year

Table 3.3 Oil and fat feedstock distribution top ten developed countries with self-sufficiency potential in 2006

Feedstock	%
Animal fats	52
Soybean oil	20
Rapeseed oil	11
Palm oil	6
Sunflower oil	5
Other vegetable oils	5

imports climbed to a record 2.5 million tons. Similarly, Pakistan, Iran, Egypt, and Bangladesh sharply increased their vegetable oil imports. In 1999, Pakistan reacted to falling vegetable oil prices with a series of increases that doubled the import duties on soybean oil and palm oil while eliminating duties on oilseeds. Pakistan also raised the import duty on soybean meal from 10% to 35% to stem the influx of Indian exports (Erickson *et al.*, 1980).

Table 3.3 shows the oil and fat feedstock distribution of the top ten developed countries with self-sufficiency potential in 2006.

3.2.1 Inedible Oil Resources

The non-edible oils such as jatropha, microalgae, neem, karanja, rubber seed, mahua, silk cotton tree, *etc.* are easily available in developing countries and are very economical comparable to edible oils.

The oils from neem (*Azardirachta indica*) and rubber (*Hevea brasiliensis*) have high free fatty acid (FFA) content. FFAs easily react with alkaline catalysts to form soap that prohibits the separation of biodiesel and glycerol. The soaps of FFAs also cause foaming in aqueous media. The resulting soaps also cause an increase in viscosity, formation of gels, and foams and make the separation of glycerol difficult (Wright *et al.*, 1944; Ma and Hanna, 1999; Demirbas, 2003a).

Vegetable oils have chemical structures different than that of petroleum-based diesel fuels. Vegetable oils containing up to three fatty acids linked to a glycerine molecule with ester linkages are called triglycerides. The fatty acids are characterized by their carbon chain length and in numbers of double bonds. There is little difference between the gross heat content among any of the vegetable oils. Their heat contents were *ca.* 88% of that of D2 fuel (Demirbas, 1998).

Vegetable oils have the potential to substitute a fraction of petroleum-based engine fuels in the near future (Demirbas, 2003b). Possible acceptable processes for converting vegetable oils into reusable products such as gasoline and diesel fuel are solvent extraction, cracking, and pyrolysis (Nagai and Seko, 2000; Bhatia *et al.*, 2003). Vegetable oil fuels are not cost competitive with petroleum-based fuels. Diesel boiling range material from plant oils is of particular interest because it has been shown to significantly reduce particulate emissions relative to petroleum diesel fuel (Giannalos *et al.*, 2002).

The advantages of biodiesel as diesel fuel are its portability, ready availability, renewability, higher combustion efficiency, lower sulfur and aromatic content (Ma and Hanna, 1999; Knothe *et al.*, 2006), higher cetane number, and higher biodegradability (Mudge and Pereira, 1999; Speidal *et al.*, 2000; Zhang *et al.*, 2003). The main advantages of biodiesel given in the literature include its domestic origin, which would reduce dependency on imported petroleum, high flash point, and inherent lubricity in the neat form (Mittelbach and Remschmidt, 2004; Knothe *et al.*, 2005).

The major disadvantages of biodiesel are its higher viscosity, lower energy content, higher cloud point and pour point, higher nitrogen oxide (NO_x) emission rates, lower engine speed and power, injector coking, engine compatibility, high price, and higher engine wear.

Biodiesel has higher cloud point and pour point compared to conventional diesel (Prakash, 1998). Neat biodiesel and biodiesel blends increase nitrogen oxide (NO_x) emissions compared with petroleum-based diesel fuel used in an unmodified diesel engine (EPA, 2002). Peak torque is less for biodiesel than petroleum diesel but occurs at lower engine speed, and generally the torque curves are flatter. Biodiesels on average decrease power by 5% compared to diesel at rated load (Demirbas, 2006).

The main commodity sources for biodiesel production from non-edible oils are plant species such as jatropha or ratanjyote or seemaikattamankku (*Jatropha curcas*), karanja or honge (*Pongamia pinnata*), nagchampa (*Calophyllum inophyllum*), rubber seed tree (*Hevca brasiliensis*), neem (*Azadirachta indica*), mahua (*Madhuca indica and Madhuca longifolia*), silk cotton tree (*Ceiba pentandra*), jojoba(*Simmondsia chinensis*), babassu tree, *Euphorbia tirucalli*, microalgae, *etc.* They are easily available in many parts of the world and are very cheap compared to edible oils in India (Karmee and Chadha, 2005).

Two major species of the genus, *Madhuca indica* and *Madhuca longifolia*, are found in India. The oil of the rubber seed tree (*Hevea brasiliensis*) is a non-edible source of biodiesel production. It is found mainly in Indonesia, Malaysia, Liberia, India, Srilanka, Sarawak, and Thailand. Rubber seed kernels (50 to 60% of seed) contain 40 to 50% of brown oil (Ramadhas *et al.*, 2004). Two major species of the genus, the oil palms *Elaeis guineensis and Elaeis oleifera*, are in Africa and Central/South America, respectively. Among vegetable oils, the price of palm oil is cheapest in palm-producing countries such as Malaysia, Indonesia, Thailand, and Korea. Neem oil is a vegetable oil pressed from the fruits and seeds of Neem (*Azadirachta indica*), an evergreen tree that is endemic to the Indian subcontinent and has been introduced to many other areas native to India and Burma, growing in tropical and semitropical regions. Jojoba oil is produced in the seed of the jojoba (*Simmondsia chinensis*) plant, a shrub native to southern Arizona, southern California, and northwestern Mexico (Wikipedia, 2007). The oil of the silk cotton tree (*Ceiba pentandra*) is a non-edible source of biodiesel production. The tree belongs to the family *Bornbacacea*. The silk cotton tree has great economic importance for both domestic and industrial uses in Nigeria. The seeds are also used as food/feed for humans and livestock in many parts of the world such as India, Tanzania, and Mozambique. *Ceiba pentendra* crude oil was extracted for 24 hours using a Soxhlet extractor with n-hexane as a solvent (Das *et al.*, 2002). Babassu tree is a species of palm tree that is a source of light yellow clear oil. There are both edible and non-edible species of babassu oils. A non-edible species of the oil is obtained from the babassu tree, which is widely grown in Brazil. The viscosity at 313.2 K and the Cetane number values of babassu oil are 3.6 cSt and 63, respectively (Srivastava and Prasad, 2000).

Fatty acid profiles of seed oils of 75 plant species having 30% or more fixed oil in their seed/kernel were examined, and *Azadirachta indica*, *Calophyllum inophyllum*, *Jatropha curcas,* and *Pongamia pinnata* were found most suitable for use as biodiesel (Azam *et al.*, 2005). The seed oil of *Jatropha* was used as a diesel fuel substitute during World War II and as blends with diesel (Foidl *et al.*, 1996; Gubitz *et al.*, 1999). Thus *Jatropha curcas* and *Pongamia pinnata (Karanja)* are most suitable for the purpose of producing renewable fuel as biodiesel (Meher *et al.*, 2006a; Meher *et al.*, 2006b). Jatropha and Karanja have high oil content (25% to 30%) (Foidl *et al.*, 1996).

From 1978 to 1996, the U.S. Department of Energy's Office of Fuels Development funded a program to develop renewable transportation fuels from algae (Sheehan *et al.*, 1998). Most current research on oil extraction is focused on microalgae to produce biodiesel from algal oil. Algal oil processes into biodiesel as easily as oil derived from land-based crops. The lipid and fatty acid contents of microalgae vary in accordance with culture conditions. All algae contain proteins, carbohydrates, lipids, and nucleic acids in varying proportions. Algal oil contains saturated and monounsaturated fatty acids. The fatty acids were determined in the algal oil in the following proportions: 36% oleic (18:1), 15% palmitic (16:0), 11% stearic (18:0), 8.4% iso-17:0, and 7.4% linoleic (18:2). The high proportion of saturated and monounsaturated fatty acids in this alga is considered optimal from a fuel quality standpoint in that fuel polymerization during combustion would be substantially less than what would occur with polyunsaturated fatty-acid-derived fuel (Sheehan *et al.*, 1998).

Oil from algae, bacteria, and fungi have also been investigated (Shay, 1993). Microalgae have been examined as a source of methyl ester diesel fuel (Nagel and Lemke, 1990), and terpenes and latexes also were studied as diesel fuels (Calvin, 1985).

Algae can grow virtually anywhere with enough sunshine. Some algae can grow in saline water. The most significant distinguishing characteristic of algal oil is in the yield and hence its biodiesel yield. According to some estimates, the yield (per acre) of oil from algae is over 200 times the yield from the best-performing plant/vegetable oils (Sheehan *et al.*, 1998). Microalgae are the fastest growing photosynthesizing organisms. They can complete an entire growing cycle every few days. Approximately 46 tons of oil/hectare/year can be produced from diatom algae. Different algae species produce different amounts of oil. Some algae produce up to 50% oil by weight. Microalgae have much faster growth rates than terrestrial crops. The per unit area yield of oil from algae is estimated to be between 5,000 and 20,000 gallons per acre per year, which is 7 to 31 times greater than the next best crop, palm oil.

3.3 Vegetable Oil Processing

Vegetable oil processing involves the extraction and processing of oils and fats from vegetable and animal resources. The oils and fats are extracted from a variety of fruits, seeds, and nuts. Natural vegetable oils and animal fats are extracted or pressed to obtain crude oil or fat. These usually contain free fatty acids, phospholipids, sterols, water, odorants, and other impurities. Even refined oils and fats contain small amounts of free fatty acids and water (Ma and Hana, 1999).

The preparation of raw materials includes husking, cleaning, crushing, and conditioning. The extraction processes are generally mechanical (boiling for fruits, pressing for seeds and nuts) or involve the use of solvent such as hexane. After boiling, the liquid oil is skimmed; after pressing, the oil is filtered; and after solvent extraction, the crude oil is separated and the solvent is evaporated and recovered. Residues are conditioned (*e.g.*, dried) and are reprocessed to yield byproducts such as animal feed. Crude oil refining includes degumming, neutralization, bleaching, deodorization, and further refining (Demirbas and Kara, 2006).

Corn, cottonseed, and peanut oil processing are similar to soybean processing, except for differences in the preparation of soybean for oil extraction. The process for soybeans typically consists of three main steps: oilseed handling/elevator operations, preparation of soybeans for solvent extraction, and oil refining.

Color-producing substances (*i.e.*, carotenoids, chlorophyll) within oil are removed by a bleaching process, which employs the use of adsorbents such as acid-activated clays. Volatile components are removed by deodorization, which uses steam injection under a high vacuum and temperature. The refined oil is then filtered and stored until used or transported.

An important step in vegetable oil purification is physical refining by removing free fatty acids present in the vegetable. This separation is carried out at a low temperature to reduce the degradation of the final products at high vacuum. The free fatty acids can be removed using stripping steam at 525 K at 2 to 3 mm Hg column top pressure.

3.3.1 Recovery of Vegetable Oils from Plants

Vegetable oils are recovered from plants by chemical extraction using solvent extracts. The most common extraction solvent is petroleum-derived hexane. Another method of vegetable oil recovery is physical extraction, which does not use solvent extracts. Supercritical carbon dioxide can also be used for extraction

Table 3.4 Properties of supercritical fluids

Compound	Boiling point (K)	Critical temperature (K)	Critical pressure (atm)
CH_3OH	337.9	513.7	78.9
CO_2	194.7	304.5	72.9
H_2O	373.2	647.6	226.8
NH_3	239.8	405.5	111.3
SiF_6	209.4	318.8	37.1
N_2O	184.2	309.7	71.4
C_6H_{14}	342.2	507.3	30.5
C_5H_{12}	309.3	469.7	34.1
C_2H_5OH	369.2	516.2	64.0

and is non-toxic (Eisenmenger *et al.*, 2005; Rajaei *et al.*, 2005). Supercritical carbon dioxide effectively extracts vegetable oils and fats (Fang *et al.*, 2007).

Supercritical fluid extraction (SFE) can reduce sample preparation time, and recovery using the methods can be equal to or better than that of the classical extraction techniques for solid and semisolid samples (Demirbas, 1991a). SFE is generally carried out in a mechanically stirring or rocking batch reactor at the solvent's critical temperature and pressure (Demirbas, 1991b). SFE is attracting a great deal of interest because the technique can considerably reduce sample preparation time and can provide analyte recovery from solid and semisolid samples that is equal to or better than that of classical extraction techniques such as Soxhlet extraction (Roselius *et al.*, 1985; Paulaitis *et al.*, 1983; Penninger *et al.*, 1985). Examples of pilot plant or commercial applications include decaffeination of coffee, removal of nicotine from tobacco, deasphalting of petroleum, extraction of oil from oilseeds, and extraction of essential oils for flavorings and perfumes (Roselius *et al.*, 1985 Schneider, 1978; Brignole, 1986). Table 3.4 shows the properties of some supercritical fluids.

3.3.2 Vegetable Oil Refining

Crude vegetable oils contain trace amounts of naturally occurring materials such as proteinaceous material, free fatty acids, and phosphatides. If physical refining is subsequently employed, it is essential to degum oils that have high phosphatide content for both economic and product quality purposes. The purpose of caustic refining is to remove free fatty acids, phosphatides, and other materials including protein meal, glycerol, carbohydrates, resins, and metals.

Crude vegetable oils obtained by oil seed processes have to be refined to remove undesirable substances. The typical oil refining process includes degumming, chemical or physical refining, bleaching, vinterization, and deodorization. Deodorization is an important step in the oil refining process. During this step, steam at 1 to 6 mm Hg pressure is injected into the oil at 490 to 550 K to eliminate free fatty acids, aldehydes, unsaturated hydrocarbons, and ketones, which cause undesirable odors and flavors in the oil (Demirbas and Kara, 2006).

The crude vegetable oil quality is very important to obtain high-quality refined oil. The oil should be efficiently degummed to remove phospholipids as well as heavy metals and bleached to remove pigments and metals.

Degummed crude oil is subjected to a further refining process. For this purpose, caustic soda (NaOH) is fed to the oil in the proper quantity to react with the free fatty acids, phosphatides, and the other impurities. Soap stock and other impurities are separated from the oil by centrifuges. Caustic soda is not completely selective in reacting with free fatty acids and phosphatides; therefore, some triglycerides are hydrolyzed and saponified (broken down and converted into soap) (Demirbas and Kara, 2006). In addition, some color reduction is also achieved by pigment removal.

3.4 The Use of Vegetable Oils as Diesel Fuel

The use of vegetable oils in diesel engines is nearly as old as the diesel engine itself. The use of vegetable oils as an alternative renewable fuel was proposed in the early 1980s (Bartholomew, 1981). The most advanced study with sunflower oil occurred in South Africa because of the oil embargo (Ma and Hanna, 1999). The first engine like the diesel engine was developed in the 1800s for fossil fuels. Famous German inventor Rudolph Diesel designed the original diesel engine to run on vegetable oil. Dr. Diesel used peanut oil to fuel one of his engines at the Paris Exposition of 1900 (Nitschke and Wilson, 1965). Because of the high temperatures created, the engine was able to run a variety of vegetable oils including hemp and peanut oil. Life for the diesel engine began in 1893 when Dr. Diesel published a paper entitled "The theory and construction of a rational heat engine". At the 1911 World's Fair in Paris, Dr. Diesel ran his engine on peanut oil and declared "the diesel engine can be fed with vegetable oils and will help considerably in the development of the agriculture of the countries which use it". One of the first uses of transesterified vegetable oil (biodiesel) was powering heavy-duty vehicles in South Africa before World War II (Demirbas, 2002a).

The first International Conference on Plant and Vegetable Oils as Fuels was held in Fargo, North Dakota in August 1982. The primary concerns discussed were the cost of the fuel, the effects of vegetable oil fuels on engine performance and durability, and fuel preparation, specifications, and additives (Ma and Hana, 1999). Oil production, oilseed processing, and extraction also were considered in this meeting (ASAE, 1982).

3.4.1 Physical and Chemical Properties of Vegetable Oils

Vegetable oils and fats are substances derived from plants that are composed of triglycerides. Oils extracted from plants have been used in many cultures since ancient times. The oily seed and nut kernels contain 20 to 60% oil. The fatty acid compositions of vegetable oils and fats are listed in Table 3.5.

As can be seen in Table 3.5, palmitic (16:0) and stearic (18:0) are the two most common saturated fatty acids, with every vegetable oil containing at least a small amount of each one. Similarly, oleic (18:1) and linoleic (18:2) were the most common unsaturated fatty acids. Many of the oils also contained some linolenic acid (18:3).

Today, the world's largest producer of soybeans is the USA, with the majority of cultivation located in the midwestern and southern USA. Soybeans must be carefully cleaned, dried, and dehulled prior to oil extraction. There are three main methods for extracting oil from soybeans. These procedures are hydraulic pressing, expeller pressing, and solvent extraction (Erickson *et al.*, 1980). Soybean oil composition is compared to other oils and normal and alternate compositions are shown and considered.

Table 3.5 Fatty acid compositions of vegetable oils and fats*

Sample	16:0	16:1	18:0	18:1	18:2	18:3	Others
Cottonseed	28.7	0	0.9	13.0	57.4	0	0
Poppyseed	12.6	0.1	4.0	22.3	60.2	0.5	0
Rapeseed	3.8	0	2.0	62.2	22.0	9.0	0
Safflowerseed	7.3	0	1.9	13.6	77.2	0	0
Sunflowerseed	6.4	0.1	2.9	17.7	72.9	0	0
Sesameseed	13.1	0	3.9	52.8	30.2	0	0
Linseed	5.1	0.3	2.5	18.9	18.1	55.1	0
Wheat grain[a]	20.6	1.0	1.1	16.6	56.0	2.9	1.8
Palm	42.6	0.3	4.4	40.5	10.1	0.2	1.1
Corn marrow	11.8	0	2.0	24.8	61.3	0	0.3
Castor[b]	1.1	0	3.1	4.9	1.3	0	89.6
Tallow	23.3	0.1	19.3	42.4	2.9	0.9	2.9
Soybean	11.9	0.3	4.1	23.2	54.2	6.3	0
Bay laurel leaf[c]	25.9	0.3	3.1	10.8	11.3	17.6	31.0
Peanut kernel[d]	11.4	0	2.4	48.3	32.0	0.9	4.0
Hazelnut kernel	4.9	0.2	2.6	83.6	8.5	0.2	0
Walnut kernel	7.2	0.2	1.9	18.5	56.0	16.2	0
Almond kernel	6.5	0.5	1.4	70.7	20.0	0	0.9
Olive kernel	5.0	0.3	1.6	74.7	17.6	0	0.8
Coconut[e]	7.8	0.1	3.0	4.4	0.8	0	65.7

* xx:y: xx number of carbon atoms; y number of double bonds

[a] Wheat grain oil contains 11.4% of 8:0 and 0.4% of 14:0 fatty acids

[b] Castor oil contains 89.6% ricinoleic acid

[c] Bay laurel oil contains 26.5% of 12:0 and 4.5% of 14:0 fatty acids

[d] Peanut kernel oil contains about 2.7% of 22:0 and 1.3% of 24:0 fatty acids

[e] Coconut oil contains about 8.9% of 8:0, 6.2% 10:0, 48.8% of 12:0, and 19.9% of 14:0 fatty acids

Diesel fuel can also be replaced by biodiesel made from vegetable oils. Biodiesel is now mainly being produced from soybean and rapeseed oils. Soybean oil is of primary interest as a biodiesel source in the USA, while many European countries use rapeseed oil, and countries with a tropical climate prefer to use coconut oil or palm oil.

Palm oil is widely grown in southeast Asia; 90% of the palm oil produced is used for food and the remaining 10% for non-food consumption, such as the production of oleo-chemicals (Leng *et al.*, 1999). An alternative use could be its conversion to liquid fuels and chemicals. Conversion of palm oil into biodiesel using methanol has been reported (Yarmo *et al.*, 1992). There are great differences between palm oil and palm kernel oil with respect to their physical and chemical characteristics. Palm oil contains mainly palmitic (16:0) and oleic (18:1) acids, the two common fatty acids, and about 50% saturated fat, while palm kernel oil contains mainly lauric acid (12:0) and more than 89% saturated fat (Demirbas, 2003c).

Rapeseed has been grown in Canada since 1936. Hundreds of years ago, Asians and Europeans used rapeseed oil in lamps. Cottonseed oil is used almost entirely as a food material. Sesame, olive, and peanut oils can be used to add flavor to a dish. Walnut oil is high-quality edible oil refined by purely physical means from quality walnuts. Poppy seeds are tiny seeds contained within the bulb of the poppy flower, also known as the opium plant (*Papaver somniferum*). Poppy seed oil is high in linoleic acid (typically 60 to 65%) and oleic acid (typically 18 to 20%) (Bajpai *et al.*, 1999).

Drying oils are vegetable oils that dry to a hard finish at normal room temperature. The polyunsaturated acid (linoleic and linolenic acids) content of a drying oil is high. Such oils are used as the basis of oil paints and in other paint and wood-finishing applications. Walnut, sunflower, safflower, dammar, linseed, poppyseed, stillingia, tang, and vernonia oils are drying oils.

Table 3.6 shows the comparisons of some fuel properties of vegetable oils with D2 fuel. The heat contents of vegetable oils are *ca.* 88% of that of D2. There is little difference between the gross heat content of any of the vegetable oils (Demirbas,

Table 3.6 Comparisons of some fuel properties of vegetable oils with D2 fuel

Fuel type	Heating value (MJ/kg)	Density (kg/m^3)	Viscosity at 300 K (mm^2 /s)	Cetane number[a]
D2 fuel	43.4	815	4.3	47.0
Sunflower oil	39.5	918	58.5	37.1
Cottonseed oil	39.6	912	50.1	48.1
Soybean oil	39.6	914	65.4	38.0
Corn oil	37.8	915	46.3	37.6
Opium poppy oil	38.9	921	56.1	–
Rapeseed oil	37.6	914	39.2	37.6

[a] Cetane number is a measure of the ignition quality of diesel fuel.

1998). The density values of vegetable oils are between 912 and 921 kg/m^3, while that of D2 fuel is 815 kg/m^3. The kinematic viscosity values of vegetable oils vary between 39.2 and 65.4 mm^2/s at 300 K. The vegetable oils are all extremely viscous, with viscosities ranging from 9 to 15 times greater than D2 fuel (Table 3.6).

Twelve crop (cottonseed, poppyseed, rapeseed, safflowerseed, sunflowerseed, soybean, corn marrow, sesameseed, linseed, castor, wheat grain, and bay laurel leaf) and five kernel (peanut, hazelnut, walnut, almond, and olive) samples used in this study were supplied from different Turkish agricultural sources. Physical analyses of the samples were carried out according to the standard test methods: ASTM D445, ASTM D613, and ASTM D524 for kinematic viscosity (KV), cetane number (CN), and carbon residue (CR), respectively. However, as too low amounts of wheat grain, bay laurel leaf, and corn marrow oils were available for determination of cetane number using the standard method, a calculated cetane number was established according to Goering *et al.* (1982). Chemical analyses of the samples were carried out according to the standard test methods: ASTM D 2015-85, ASTM D5453-93, ASTM D482-91, AOCS CD1-25, and AOCS CD3 for higher heating value (HHV), sulfur content, ash content, iodine value, and saponification value, respectively. The other standard test methods for fuel properties are presented in Table 3.7. Fatty acid compositions of oil samples were determined by gas chromatographic (GC) analysis. The oil samples were saponified for 3.5 h and 338 K with 0.5 N methanolic KOH to liberate the fatty acids present as their esters. After acidification of the saponified solutions with 1.5 N HCl acid, acids were then weighed and methylated with diazomethane according to the method of Schelenk and Gellerman (1960). The methyl esters of the fatty acids were analyzed by GC (Hewlett-Packard 5790) on a 12 m long and 0.2 mm inside diameter capillary column coated with Carbowax PEG 20. The detector was a FID. Helium was used as carrier gas. The flame ionization detector temperature was 500 K.

The oven temperature was kept at 450 K for 25 min. After that, the oven was heated with a heat ratio of 5 K/min to 495 K. Spectra of methyl esters were recorded

Table 3.7 Determination of physical and chemical properties using standard test methods

Property	Symbol	Standard method	Unit
Density	d	ASTM D4052-91	g/ml
Iodine value	IV	AOCS CD1-25 1993	centigram I/g oil
Saponification value	SV	AOCS CD3 1993	mg KOH/g oil
Higher heating value	HHV	ASTM D2015-85	MJ/kg
Cloud point	CP	ASTM D2500-91	K
Pour point	PP	ASTM D97-93	K
Flash point	FP	ASTM D93-94	K
Cetane number	CN	ASTM D613	–
Kinematic viscosity	KV	ASTM D445	mm^2/s at 311 K
Sulfur content	SC	ASTM D5453-93	wt.%
Carbon residue	CR	ASTM D 524	wt.%
Ash content	AC	ASTM D482-91	wt.%

Table 3.8 Physical and chemical properties of oil samples

Vegetable oil	KV	CR	CN	HHV	AC	SC	IV	SV
Cottonseed	33.7	0.25	33.7	39.4	0.02	0.01	113.20	207.71
Poppyseed	42.4	0.25	36.7	39.6	0.02	0.01	116.83	196.82
Rapeseed	37.3	0.31	37.5	39.7	0.006	0.01	108.05	197.07
Safflowerseed	31.6	0.26	42.0	39.5	0.007	0.01	139.83	190.23
Sunflowerseed	34.4	0.28	36.7	39.6	0.01	0.01	132.32	191.70
Sesameseed	36.0	0.25	40.4	39.4	0.002	0.01	91.76	210.34
Linseed	28.0	0.24	27.6	39.3	0.01	0.01	156.74	188.71
Wheat grain	32.6	0.23	35.2	39.3	0.02	0.02	120.96	205.68
Corn marrow	35.1	0.22	37.5	39.6	0.01	0.01	119.41	194.14
Castor	29.7	0.21	42.3	37.4	0.01	0.01	88.72	202.71
Soybean	33.1	0.24	38.1	39.6	0.006	0.01	69.82	220.78
Bay laurel leaf	23.2	0.20	33.6	39.3	0.03	0.02	105.15	220.62
Peanut kernel	40.0	0.22	34.6	39.5	0.02	0.01	119.55	199.80
Hazelnut kernel	24.0	0.21	52.9	39.8	0.01	0.02	98.62	197.63
Walnut kernel	36.8	0.24	33.6	39.6	0.02	0.02	135.24	190.82
Almond kernel	34.2	0.22	34.5	39.8	0.01	0.01	102.35	197.56
Olive kernel	29.4	0.23	49.3	39.7	0.008	0.02	100.16	196.83

with a VG 70-250-SE mass spectrometer with double focusing. Ionization was carried out at 70 eV. The mass spectrometer was fitted to the gas chromatograph by means of a capillary glass jet separator.

Table 3.8 lists the physical and chemical properties of the oil samples. Viscosity values (KVs) of the oil samples range from 23.2 to 42.4 mm^2/s at 311 K. The vegetable oils are all extremely viscous, with viscosities ranging from 10 to 20 times that of the ASTM upper limit given for diesel fuels (2.7 mm^2/s).

Higher heating values (HHVs) of the oil samples range from 39.3 to 39.8 MJ/kg. Castor oil has an exceptional HHV (37.4 MJ/kg). The oxygen content of castor oil is higher than that of vegetable oils due to the ricinoleic acid in its structure. Ricinoleic acid is the only 18:1 fatty acid that contains a hydroxyl group. Because castor oil contains a hydroxyl group, its HHV is lower than the others.

The cetane numbers of the oil samples range from 27.6 to 52.9. The iodine values (IVs) and saponification values (SVs) of the oil samples range from 69.82 to 156.74 and from 188.71 to 220.78, respectively.

The SV of an oil decreases with increase of its molecular weight. The percentages of C and H in an oil decrease with an increase in molecular weight. The increase in SV results in a decrease in the heat content of an oil. The increase in IV (*i.e.*, carbon-carbon double bond, –C=C–, content of oil) results in a decrease in the heat content of an oil. The heat content of the oil depends on the SV and IV. Therefore, for calculation of the HHVs (MJ/kg) of oil samples, Eq. (3.1) was proposed (Demirbas, 1998).

$$HHV = 49.43 - 0.041\,(SV) - 0.015\,(IV) \tag{3.1}$$

Thus the heating values of the vegetable oils can be calculated using their SVs and IVs obtained non-calorimetrically from simple chemical analyses.

The carbon residue (CR), sulfur content (SC), and ash content (AC) of the oil samples range from 0.20 to 0.31 wt.%, from 0.01 to 0.02 wt.%, and from 0.002 to 0.02 wt.%, respectively (Table 3.8).

The increase in heat content results from a high increase in the number of carbons and hydrogens, as well as an increase in the ratio of these elements relative to oxygen. A decrease in heat content is the result of fewer hydrogen atoms (*i.e.*, greater unsaturation) in the fuel molecule. An examination of data obtained for a great many compounds has shown that the HHV of an aliphatic hydrocarbon agrees rather closely with that calculated by assuming a certain characteristic contribution from each structural unit (Morrison and Boyd, 1983). For open-chain alkanes, each methylene group, $-CH_2-$, contributes very close to 46,956 kJ/kg.

3.4.2 Direct Use of Vegetable Oils in Diesel Engines

The use of vegetable oils as an alternative renewable fuel competing with petroleum was proposed in the early 1980s. The advantages of vegetable oils as diesel fuel are as follows (Demirbas, 2003b):

- Portability
- Ready availability
- Renewability
- Higher heat content (about 88% of D2 fuel)
- Lower sulfur content
- Lower aromatic content
- Biodegradability

Full combustion of a fuel requires in existence the amount of stoichiometric oxygen. However, the amount of stochiometric oxygen generally is not enough for full combustion because the fuel is not oxygenated. The structural oxygen content of fuel increases the combustion efficiency of the fuel due to increased mixing of oxygen with the fuel during combustion. For these reasons, the combustion efficiency and cetane number of vegetable oil are higher than those of diesel fuel, and the combustion efficiency of methanol and ethanol is higher than that of gasoline.

The disadvantages of vegetable oils as diesel fuel are (Pryde, 1983):

- Higher viscosity
- Lower volatility
- The reactivity of unsaturated hydrocarbon chains

Problems appear only after the engine has been operating on vegetable oils for longer periods of time, especially with direct-injection engines. The problems include (a) coking and trumpet formation on the injectors to such an extent that fuel atomization does not occur properly or is even prevented as a result of plugged orifices, (b) carbon deposits, (c) oil ring sticking, and (d) thickening and gelling of the lubricating oil as a result of contamination by the vegetable oils (Ma and Hana, 1999).

Among the renewable resources for the production of alternative fuels, triglycerides have attracted much attention as alternative diesel engine fuels (Shay, 1993). However, the direct use of vegetable oils and/or oil blends is generally considered to be unsatisfactory and impractical for both direct injection and indirect type diesel engines because of their high viscosities and low volatilities, injector coking and trumpet formation on the injectors, higher level of carbon deposits, oil ring sticking, and thickening and gelling of the engine lubricant oil, acid composition (the reactivity of unsaturated hydrocarbon chains), and free fatty acid content (Ma an Hanna, 1999; Darnoko and Cheryan, 2000; Srivastava and Prasad, 2000; Komers *et al.* 2001). Consequently, different methods have been considered to reduce the viscosity of vegetable oils such as dilution, microemulsification, pyrolysis, catalytic cracking, and transesterification. Methods based on pyrolysis (Alencar *et al.* 1983; Dykstra *et al.*, 1988; Agra *et al.*, 1992; Adjaye *et al.*, 1995, 1996; Shay, 1993; Dandik and Aksoy, 1998; Bhatia *et al.*, 1999; Lima *et al.*, 2004) and microemulsification (Billaud *et al.*, 1995) have been studied but are not entirely satisfactory.

Vegetable oils can be used as fuels for diesel engines, but their viscosities are much higher than that of common diesel fuel and require modifications of the engines (Kerschbaum and Rinke, 2004). Vegetable oils could only replace a very small fraction of transport fuel. Different methods of reducing the high viscosity of vegetable oils have been considered:

1. Dilution of 25 parts of vegetable oil with 75 parts of diesel fuel
2. Microemulsions with short-chain alcohols such as ethanol or methanol
3. Transesterification with ethanol or methanol, which produces biodiesel
4. Pyrolysis and catalytic cracking, which produces alkanes, cycloalkanes, alkenes, and alkylbenzenes.

3.4.2.1 Dilution of Oils

Dilution of oils with solvents and microemulsions of vegetable oils lowers the viscosity and mitigates some engine performance problems such as injector coking and carbon deposits, *etc.* To dilute vegetable oils the addition of 4% ethanol to the oils increases the brake thermal efficiency, brake torque, and brake power while decreasing brake-specific fuel consumption. Since the boiling point of ethanol is less than that of vegetable oils, it could assist in the development of the combustion process through an unburned blend spray (Bilgin *et al.*, 2002).

The viscosity of oil can be lowered by blending with pure ethanol. Twenty-five parts of sunflower oil and 75 parts of diesel were blended as diesel fuel (Ziejewski *et al.*, 1986). The viscosity was 4.88 cSt at 313 K, while the maximum specified ASTM value was 4.0cSt at 313 K. This mixture was not suitable for long-term use in a direct injection engine. Another study was conducted using the dilution technique on the same frying oil (Karaosmonoglu, 1999).

The addition of 4% ethanol to D2 fuel increases the brake thermal efficiency, brake torque, and brake power while decreasing brake-specific fuel consumption. Since the boiling point of ethanol is less than that of D2 fuel, it could assist the development of the combustion process through an unburned blend spray (Bilgin *et al.*, 2002).

3.4.2.2 Microemulsion of Oils

To reduce the high viscosity of vegetable oils, microemulsions with immiscible liquids such as methanol and ethanol and ionic or non-ionic amphiphiles have been studied (Billaud *et al.*, 1995). The short engine performances of both ionic and non-ionic microemulsions of ethanol in soybean oil were nearly as good as that of D2 fuel (Goering *et al.*, 1982).

To solve the problem of the high viscosity of vegetable oils, microemulsions with solvents such as methanol, ethanol, and 1-butanol have been studied. All microemulsions with butanol, hexanol, and octanol met the maximum viscosity requirement for D2 fuel. The 2-octanol was found to be an effective amphiphile in the micellar solubilization of methanol in triolein and soybean oil (Schwab *et al.*, 1987; Ma and Hanna, 1999).

Ziejewski *et al.* (1984) prepared an emulsion of 53% (vol) alkali-refined and winterized sunflower oil, 13.3% (vol) 190-proof ethanol, and 33.4% (vol) 1-butanol. This non-ionic emulsion had a viscosity of 6.31 cSt at 313 K, a cetane number of 25, and an ash content of less than 0.01%. Lower viscosities and better spray patterns (more even) were observed with an increase of 1-butanol. In a 200-h laboratory screening endurance test, no significant deteriorations in performance were observed, but irregular injector needle sticking, heavy carbon deposits, incomplete combustion, and an increase in lubricating oil viscosity were reported (Ma and Hanna, 1999).

A microemulsion prepared by blending soybean oil, methanol, 2-octanol, and cetane improver in the ratio of 52.7:13.3:33.3:1.0 also passed the 200-h EMA test (Goering, 1984). Schwab *et al.* (1987) used the ternary phase equilibrium diagram and the plot of viscosity versus solvent fraction to determine the emulsified fuel formulations. All microemulsions with butanol, hexanol, and octanol met the maximum viscosity requirement for D2. The 2-octanol was an effective amphiphile in the micellar solubilization of methanol in triolein and soybean oil. Methanol was often used due to economic advantage over ethanol.

3.4.2.3 Transesterification of Oils and Fats

Among all these alternatives, transesterification seems to be the best choice as the physical characteristics of fatty acid esters (biodiesel) are very close to those of diesel fuel and the process is relatively simple. In the esterification of an acid, an alcohol acts as a nucleophilic reagent; in the hydrolysis of an ester, an alcohol is displaced by a nucleophilic reagent. This alcoholysis (cleavage by an alcohol) of an ester is called transesterification (Gunstone and Hamilton, 2001).

Transesterified vegetable oils have proven to be a viable alternative diesel engine fuel with characteristics similar to those of diesel fuel. The transesterification reaction proceeds with catalyst or any unused catalyst by using primary or secondary monohydric aliphatic alcohols having between one and eight carbon atoms as follows:

$$\text{Triglycerides} + \text{Monohydric alcohol} \leftrightarrows \text{Glycerine} + \text{Monoalkyl esters.} \quad (3.2)$$

Transesterification is catalyzed by a base (usually alkoxide ion) or acid (H_2SO_4 or dry HCl). The transesterification is an equilibrium reaction. To shift the equilibrium to the right, it is necessary to use a large excess of the alcohol or else to remove one of the products from the reaction mixture. Furthermore, the methyl or ethyl esters of fatty acids can be burned directly in unmodified diesel engines, with very low deposit formation. Although short-term tests using neat vegetable oil showed promising results, longer tests led to injector coking, more engine deposits, ring sticking, and thickening of the engine lubricant. These experiences led to the use of modified vegetable oil as a fuel.

Technical aspects of biodiesel are close to petroleum diesel, such as physical and chemical characteristics of methyl esters related to its performance in compression ignition engines (Saucedo, 2001). Compared with transesterification, pyrolysis has more advantages. The liquid fuel produced from pyrolysis has similar chemical components to conventional petroleum diesel fuel (Zhenyi *et al.*, 2004).

3.4.2.4 Pyrolysis and Catalytic Cracking

Pyrolysis is the conversion of one substance into another by means of heat or by heat with the aid of a catalyst (Sonntag, 1979). It involves heating in the absence of air or oxygen and cleavage of chemical bonds to yield small molecules (Weisz *et al.*, 1979). The pyrolyzed material can be vegetable oils, animal fats, natural fatty acids, and methyl esters of fatty acids.

Soybean oil was thermally decomposed and distilled in air and nitrogen sparged with a standard ASTM distillation apparatus (Niehaus *et al.*, 1986; Schwab *et al.*, 1988). The main components were alkanes and alkenes, which accounted for *ca.* 60% of the total weight. Carboxylic acids accounted for another 9.6 to 16.1%.

Catalytic cracking of vegetable oils to produce biofuels has been studied (Pioch *et al.*, 1993). Copra oil and palm oil stearin were cracked over a standard petroleum catalyst SiO_2/Al_2O_3 at 723 K to produce gases, liquids, and solids with lower molecular weights. The condensed organic phase was fractionated to produce biogasoline and biodiesel fuels.

3.5 New Engine Fuels from Vegetable Oils

Vegetable oils as alternative fuels can be used for diesel engines. Due to the rapid decline in crude oil reserves, the use of vegetable oils as diesel fuels is again being promoted in many countries. The effect of coconut oil as a diesel fuel alternative or as direct fuel blends is being investigated using a single-cylinder, direct-injection diesel engine (Machacon *et al.*, 2001).

Vegetable oils have the potential to substitute a fraction of petroleum distillates and petroleum-based petrochemicals in the near future. Vegetable oil fuels are not petroleum-competitive fuels because they are more expensive than petroleum fuels. However, with recent increases in petroleum prices and uncertainties concerning petroleum availability, there is renewed interest in using vegetable oils in diesel engines.

The soaps obtained from vegetable oils can be pyrolyzed into hydrocarbon-rich products (Demirbas, 2002b). Pyrolysis of Na-soaps may be carried out on vegetable oil products as follows:

$$2RCOONa \rightarrow R{-}R + Na_2CO_3 + CO. \quad (3.3)$$

The soaps obtained from the vegetable oils can be pyrolyzed into hydrocarbon-rich products according to Eq. (3.3) with higher yields at lower temperatures (Demirbas, 2002b). Table 3.9 shows the yields of pyrolysis products from used sunflower oil sodium soaps at different temperatures. These findings are in general agreement with results given in the literature (Barsic and Humke, 1981).

Table 3.9 Yields of pyrolysis products from sunflower oil sodium soaps at different temperatures (% by weight)

400 K	450 K	500 K	520 K	550 K	570 K	590 K	610 K
2.8	8.4	29.0	45.4	62.4	84.6	92.7	97.5

3.5.1 Pyrolysis of Vegetable Oils and Fats

Pyrolysis/cracking, defined as the cleavage to smaller molecules by thermal energy, of vegetable oils over petroleum catalysts has been investigated (Madras *et al.*, 2004). Because of the possibility of producing triglycerides in a wide variety of products by high-temperature pyrolysis reactions, many investigators have studied the pyrolysis of triglycerides to obtain products (liquid, gas, and solid) suitable for fuel under different reaction conditions with and without catalyst (Dykstra *et al.*, 1988; Agra *et al.*, 1992; Adjaye *et al.*, 1995, 1996; Dandik and Aksoy, 1998; Bhatia *et al.*, 1999; Lima *et al.*, 2003; Bhatia *et al.*, 2003).

Pyrolysis is difficult to precisely define, especially when applied to biomass. The older literature generally equates pyrolysis with carbonization, in which the principal product is a solid char by very slow pyrolysis. Currently, the term pyrolysis often describes processes in which oils are preferred products. The time frame for pyrolysis is much faster for the latter process (Mohan *et al.*, 2006).

Pyrolysis is the thermal degradation of vegetable oils by heat in the absence of oxygen, which results in the production of alkanes, alkenes, alkadienes, carboxylic acids, aromatics, and small amounts of gaseous products. Depending on the operating conditions, the pyrolysis process can be divided into three subclasses: conventional pyrolysis, fast pyrolysis, and flash pyrolysis.

Pyrolysis of triglycerides has been investigated for more than 100 years, especially in areas of the world that lack deposits of petroleum (Zhenyi *et al.*, 2004). Pyrolysis of used sunflower oil was carried out in a reactor equipped with a fractionating packed column at 673 and 693 K in the presence of sodium carbonate (1, 5, 10, and 20% based on oil weight) as a catalyst. The conversion of oil was high (42 to 83 wt.%) and the product distribution depended strongly on the reaction temperature, residence times, and catalyst content. The pyrolysis products consisted of gas and liquid hydrocarbons, carboxylic acids, CO, CO_2, H_2, and water (Dandik and Aksoy, 1996).

There are several parameters that should be controlled during experiments such as temperature, residence times, and catalyst content. The three vegetable oils (soybean, palm, and castor oils) were pyrolyzed to obtain light fuel products at 503 to 673 K (Lima *et al.*, 2004). These results show that soybean, palm, and castor oils present a similar behavior depending on the pyrolysis temperature range. On the other hand, palm oil reacts in a lower temperature range with a higher yield in the heavy fraction (Lima *et al.*, 2004). A short pyrolysis time (less than 10 s) leads to a high amount of alkanes, alkenes, and aldehydes instead of carboxylic acids. On the other hand, higher temperature and long pyrolysis times do not favor an pyrolysis of this material. In this case, a process like desorption becomes more likely than the pyrolytic process. The liquid products can be improved by deoxygenation in order to obtain an enriched hydrocarbon diesel-like fuel (Lima *et al.*, 2004; Fortes and Baugh, 1999). The parameters of the pyrolysis systematically affect the pyrolytic process.

Increasing Na_2CO_3 content and temperature increases the formation of liquid hydrocarbon and gas products and decreases the formation of aqueous-phase, acid-phase, and coke–residual oil. The highest C_5–C_{11} yield (36.4%) was obtained using 10% Na_2CO_3 and a packed column of 180 mm at 693 K. The use of a packed column increased the residence times of the primer pyrolysis products in the reactor and packed column by the fractionating of the products, which caused the additional catalytic and thermal reactions in the reaction system and increased the content of liquid hydrocarbons in the gasoline boiling range (Dandik and Aksoy, 1996).

The HHV of pyrolysis oil from vegetable oils is pretty high. The HHV of pyrolysis oil from rapeseed (38.4 MJ/kg) is slightly lower than that of gasoline (47 MJ/kg), diesel fuel (43 MJ/kg), or petroleum (42 MJ/kg) but higher than coal (32–37 MJ/kg) (Sensoz and Angin, 2000).

3.5.1.1 Catalytic Pyrolysis of Sunflower Oil

Table 3.10 shows the yields of $ZnCl_2$ catalytic pyrolysis from sunflower oil at different temperatures (Demirbas, 2003c). The yield of conversion into products of $ZnCl_2$ catalytic pyrolysis from sunflower oil increased with increases in reaction temperature. The yield of conversion into products from sunflower oil reached a maximum of 78.3% at 660 K. The decrease in yield of conversion could probably be due to higher coke and gas formation at pyrolysis temperatures higher than 660 K. As more coke deposited on the catalyst's surface, the effect of pyrolysis diminished. The gasoline content reached a maximum (35.8% of the conversion products) at 660 K. The aromatic and gas oil contents of conversion products showed a similar trend (Demirbas, 2003c).

Table 3.10 Yields of $ZnCl_2$ catalytic pyrolysis from sunflower oil (SFO) at different temperatures

Temperature (K)	610	630	650	660	670	690
Conversion, wt.% of SFO	35.6	60.7	71.5	78.3	74.9	68.4
Gaseous product, wt.%	3.4	5.1	6.4	7.0	8.9	10.6
Aromatic content, wt.%	8.5	9.3	9.0	9.6	8.2	8.8
Gasoline content, wt.%	28.6	30.4	29.4	35.8	32.7	29.3
Gas oil content, wt.%	6.6	7.3	8.4	10.7	8.6	7.9
Coke residue, wt.%	0.2	0.3	0.4	0.5	2.4	6.8
Water formation, wt.%	3.4	3.7	4.1	4.5	4.1	3.8
Unidentified, wt.%	49.3	43.9	42.3	31.9	35.1	32.8

3.5.2 Cracking of Vegetable Oils

The thermal degradation of aliphatic long-chain compounds is known as cracking. Higher-molecular-weight molecules generally convert into smaller-molecular-weight molecules by the cracking process. Large alkane molecules are converted into smaller alkanes and some hydrogen in the cracking process. Smaller hydrocarbons can be obtained by a hydrocracking process. The hydrocracking is carried out in the presence of a catalyst and hydrogen, at high pressure and at much lower temperatures (525 to 725 K).

Higher boiling petroleum fractions (typically gas oil) are obtained from silica-alumina catalytic cracking at 725 to 825 K and under lower pressure. The catalytic cracking not only increases the yield of gasoline by breaking large molecules into smaller ones but also improves the quality of the gasoline: this process involves carbocations (the carbocation, a group of atoms that contains a carbon atom bearing only six electrons).

Palm oil stearin and copra oil was subjected to conversion over different catalysts like silica-alumina and zeolite Pioch *et al.*, 1993). It was found that the conversion of palm and copra oil was 84 wt.% and 74 wt.%, respectively. The silica-alumina catalyst was highly selective for obtaining aliphatic hydrocarbons, mainly in the kerosene boiling point range (Katikaneni *et al.*, 1995). The organic liquid products obtained with a silica-alumina catalyst contained between 4 and 31 wt.% aliphatic hydrocarbons and 14 and 58 wt.% aromatic hydrocarbons. The conversion was high and ranged between 81 and 99 wt.%. Silica-alumina catalysts are suitable for converting vegetable oils into aliphatic hydrocarbons. Zinc chloride catalyst, as a Lewis acid, contributed to hydrogen transfer reactions and the formation of hydrocarbons in the liquid phase. Palm oil was converted into hydrocarbons using a shape selective zeolite catalyst (Leng *et al.*, 1999). Palm oil can be converted into gasoline, diesel and kerosene, light gases, coke and water with a yield of 70 wt.%. The maximum yield of gasoline-range hydrocarbons was 40 wt.% of the total product.

Palm oil has been cracked at atmospheric pressure at a reaction temperature of 723 K to produce biofuel in a fixed-bed microreactor. The reaction was carried out over microporous HZSM-5 zeolite, mesoporous MCM-41, and composite micro-mesoporous zeolite as catalysts. The products obtained were gas, organic liquid, water, and coke. The organic liquid product was composed of hydrocarbons corresponding to the gasoline, kerosene, and diesel boiling point ranges. The maximum conversion of palm oil, 99 wt.%, and gasoline yield of 48 wt.% was obtained with composite micromesoporous zeolite (Sang *et al.*, 2003). Table 3.9 presents the conversion of palm oil over HZSM-5 with different Si/Al ratios of catalyst by catalytic cracking. The gasoline yield increased with an increase in the Si/Al ratio due to the decrease in the secondary cracking reactions and the drop in the yield of gaseous products (Sang *et al.*, 2003). The vegetable oils could be converted into liquid products containing gasoline-boiling-range hydrocarbons. The results show that the product compositions are affected by catalyst content and temperature.

3.5.3 Pyrolysis Mechanisms of Vegetable Oils

Pyrolysis involves thermal degradation of biomass by heat in the absence of oxygen, which results in the production of charcoal (solid), bio-oil (liquid), and fuel gaseous products. The pyrolysis of biomass has been studied with the ultimate objective of recovering a biofuel with a medium-low heating value (Maschio *et al.*, 1992; Barth, 1999; Bridgwater *et al.*, 1999).

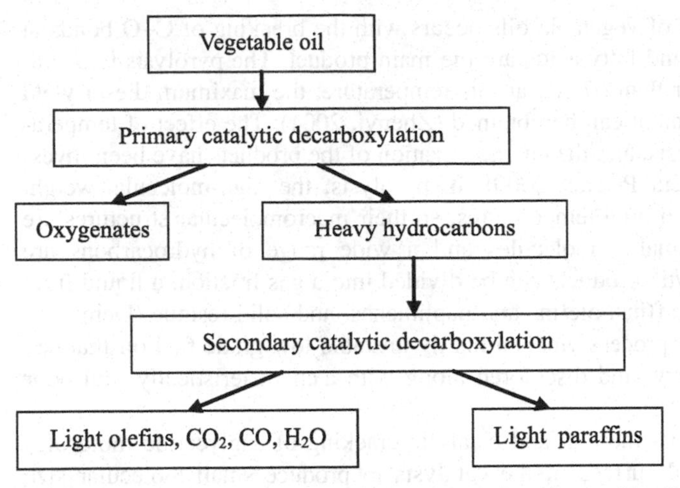

Fig. 3.3 Reaction pathway of catalytic decarboxylation of vegetable oils

Soybean, rapeseed, sunflower, and palm oils are the most studied for the preparation of bio-oil. In one study, the viscosity of the distillate was 10.2 mm^2/s at 311 K, which is higher than the ASTM specification for D2 fuel (1.9 to 4.1 mm^2/s) but considerably below that of soybean oil (32.6 mm^2/s). Cottonseed oil used in the cooking process was decomposed with Na_2CO_3 as catalyst at 725 K to give a pyrolyzate containing mainly C8–20 alkanes (69.6%) in addition to alkenes and aromatics. The pyrolyzate had a lower viscosity, pour point, and flash point than D2 fuel and equivalent heating values (Bala, 2005).

A mechanism for catalytic decarboxylation of vegetable oils is presented in Fig. 3.3. Vegetable oils contain mainly palmitic, stearic, oleic, and linoleic acids. These fatty acids underwent various reactions, resulting in the formation of different types of hydrocarbons.

The variety of reaction paths and intermediates makes it difficult to describe the reaction mechanism. In addition, the multiplicity of possible reactions of mixed triglycerides makes pyrolysis reactions more complicated (Zhenyi *et al.*, 2004). Generally, thermal decomposition of triglycerides proceeds through either a free-radical or carbonium ion mechanism (Srivastava and Prasad, 2000). Vegetable oil is converted into lower molecular products by two simultaneous reactions: cracking and condensation. The heavy hydrocarbons produced from primary and secondary deoxygenation and cracking yield light olefins and light paraffins, water, carbon dioxide, and carbon monoxide. Hydrocarbon formation can be identified as deoxygenation, cracking, and aromatization with H-transfer. Deoxygenation can take place via decarboxylation and dehydration (Chang and Silvestri, 1977)

The distribution of pyrolysis products depends on the dynamics and kinetic control of different reactions. The maximum gasoline fraction can be obtained under appropriate reaction conditions. Thermodynamic calculation shows that the initial decomposition of vegetable oils occurs with the breaking of C–O bonds at lower temperatures, and fatty acids are the main product. The pyrolysis temperature should be higher than 675 K; at this temperature, the maximum diesel yield with high oxygen content can be obtained (Zhenyi, 2004). The effect of temperature, the use of catalysts, and the characterization of the products have been investigated (Srivastava and Prasad, 2000). In pyrolysis, the high-molecular-weight materials are heated to high temperatures, so their macromolecular structures are broken down into smaller molecules and a wide range of hydrocarbons are formed. These pyrolytic products can be divided into a gas fraction, a liquid fraction consisting of paraffins, olefins and naphthenes, and solid residue (Demirbas, 2004a). The cracking process yields a highly unstable low-grade fuel oil that can be acid-corrosive, tarry, and discolored along with a characteristically foul odor (Demirbas, 2004b)

It was proposed that thermal and catalytic cracking of triglyceride molecules occurs at the external surface of the catalysts to produce small molecular size components, comprised mainly of heavy liquid hydrocarbons and oxygenates (Leng *et al.*, 1999). In general, it is assumed that the reactions occur predominantly within the internal pore structure of a zeolite catalyst.

The catalyst acidity and pore size affect the formation of aromatic and aliphatic hydrocarbons. Hydrogen transfer reactions, essential for hydrocarbon formation, are known to increase with catalyst acidity. The high acid density of $ZnCl_2$ catalysts contribute greatly to high amounts of hydrocarbons in the liquid product.

3.6 Gasoline-rich Liquid from Sunflower Oil by Alumina Catalytic Pyrolysis

Recycling and rerefining are the applicable processes for upgrading of vegetable oils by converting them into reusable products such as gasoline and diesel fuel. Possible acceptable processes are transesterification, cracking, and pyrolysis (Nagai and Seko, 2000; He *et al.*, 2007). Samples of sunflower seed oil were used in the experiments. The sunflower oils were obtained from commercial sources and used without further purification. Aluminum oxide (Al_2O_3, also known as alumina) was obtained from bauxite by a caustic leach method. The catalyst was treated with 10% sodium hydroxide solution before being used in the pyrolysis. 2.5 g of NaOH and 25 g alumina (Al_2O_3) were added to 250 ml of deionized water and stirred into a water bath for 45 min. 0.5 g of $AlCl_3$ was slowly added to the mixture and stirred vigorously for 30 min. The solid material was thoroughly washed, filtered, dried at room temperature overnight, and then calcined at 850 K for 6 h.

The pyrolysis experiments were performed in a laboratory-scale apparatus. The main element of this device was a vertical cylindrical reactor of stainless steel (127.0 mm height, 17.0 mm inner diameter, and 25.0 mm outer diameter) inserted vertically into an electrically heated furnace and provided with an electrical

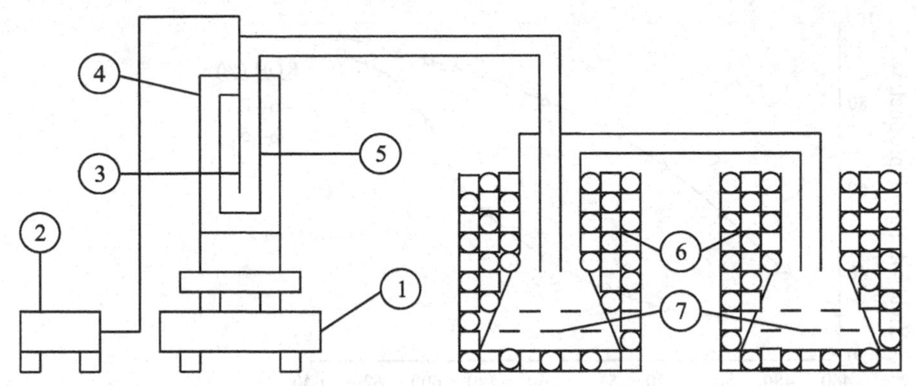

Fig. 3.4 Simplified device of catalytic pyrolysis. (1) Digital balance, (2) Temperature transmitter, (3) Thermocouple, (4) External heater, (5) Steel reaction vessel, (6) Ice bath, and (7) Collection vessels for liquid products

heating system power source. Figure 3.4 shows the simplified device of the catalytic pyrolysis. The average heating rate was 5 K/min.

Heat to the vertical cylindrical reactor was supplied from an external heater and the power was adjusted to give an appropriate heat-up time. A simple thermocouple (NiCr – Constantan) or a 360° thermometer with mercury was placed directly in the pyrolysis medium. For each run, the heater was started at 298 K and terminated when the desired temperature was reached. The sunflower seed oil samples were treated with 3% sodium hydroxide solutions in a separatory funnel and then washed with water before pyrolysis. The catalyst (1, 3, 4, and 5% by weight of the used sample) was used in the pyrolysis experiments. In addition, the sunflower seed oil samples were pyrolyzed in catalytic runs with 2 and 5% potassium hydroxide. The pyrolysis products were collected within three different groups as condensable liquid products, non-condensable gaseous products, and solid residue. The liquid product was collected in two glass traps with a cooled ice–salt mixture and ice, respectively. The gas products were trapped over a saturated solution of NaCl in a gas holder.

Figure 3.5 shows the plots for yield of liquid products from pyrolysis of the sunflower oil at different temperatures in the presence of KOH. The nominal pyrolysis time was 30 min. The yields of liquid products increase with increasing temperature and amounts of KOH. The yield sharply increases between 580 and 610 K and then reaches a plateau value with a 2% KOH run. Qualitative observations show that the pyrolytic liquid products from the runs with KOH are highly viscous by comparing the waste-cooking sunflower oil. The repolymerization degree of the pyrolytic liquid products increases with increasing temperature.

Figure 3.6 shows the plots for the yields of liquid products from pyrolysis of sunflower oil at different temperatures in the presence of aluminum oxide catalyst. The catalyst was treated with 10% sodium hydroxide solution before use in the pyrolysis. Particle size of the catalyst was 80 to 120 mesh. The yields of liquid products generally increase with increasing temperature and the percent of catalyst.

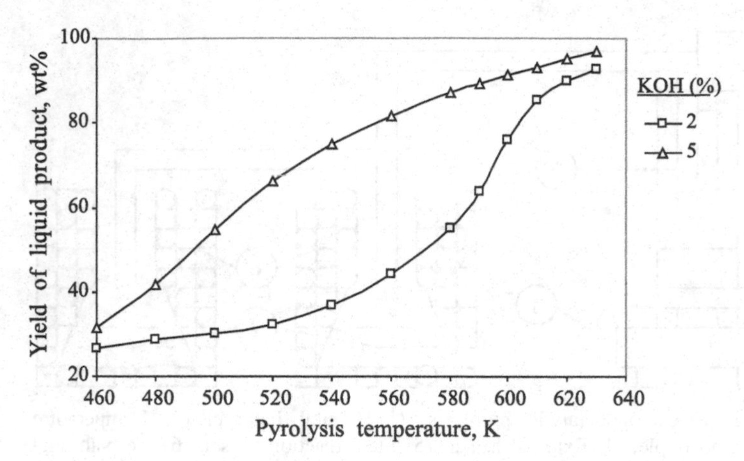

Fig. 3.5 Yield of liquid products from pyrolysis of sunflower oil at different temperatures in presence of KOH. Pyrolysis time: 30 min

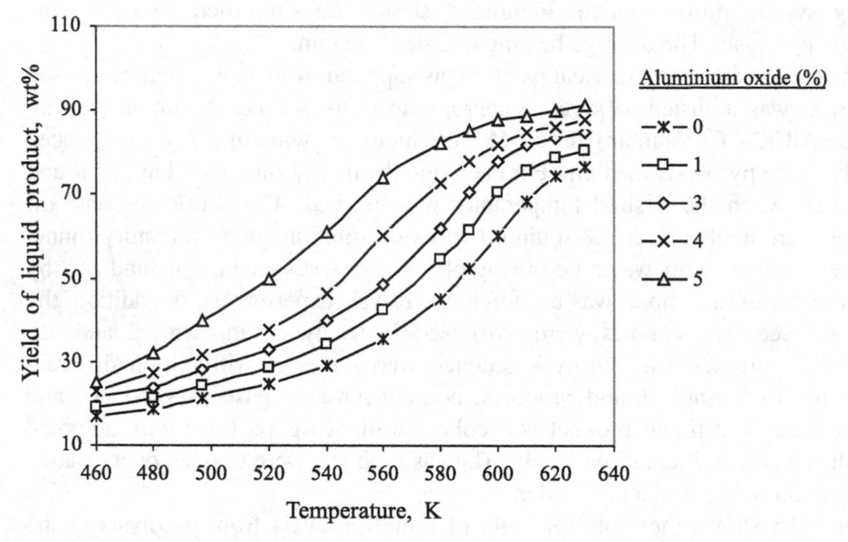

Fig. 3.6 Yield of liquid products from pyrolysis of sunflower oil at different temperatures in the presence of aluminum oxide. Pyrolysis time: 30 min. Particle size: 80–120 mesh

Figure 3.6 shows how the yield of liquid product from sunflower oil sharply increases between 500 and 680 K in 5% catalytic runs. The yields from non-catalytic runs were 22.1 and 76.8% at 500 and 630 K, respectively. The yields from 5% catalytic runs were 39.8 and 91.4% at 500 and 630 K, respectively. The yields of liquid products reach plateau values between 600 and 630 K.

The liquid products from pyrolysis of used samples have gasolinelike fractions. Table 3.11 shows the average gasoline percentages of liquid products from pyrolysis of sunflower seed oil at different temperatures in the presence of

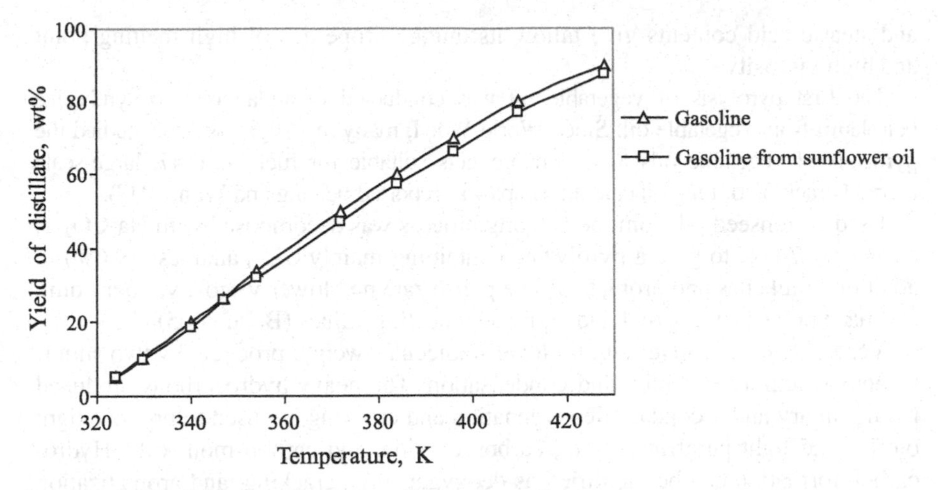

Fig. 3.7 Distillation of gasoline and sunflower oil gasoline

Table 3.11 Average gasoline percentages of liquid products from pyrolysis of sunflower seed oil at different temperatures in the presence of aluminum oxide

Al_2O_3 (%)	560 K	580 K	600 K	620 K	630 K
0	5.7	9.6	11.8	13.7	16.8
1	12.5	19.3	23.7	27.9	32.5
3	17.3	23.5	28.4	32.8	38.2
4	24.9	29.4	33.9	38.7	47.4
5	33.5	39.6	42.5	48.1	53.8

sodium-hydroxide-treated aluminum oxide. As seen from Table 3.11, the properties of liquid products obtained from catalytic pyrolysis are similar to those of gasoline. The highest yields of gasoline were 53.8% for the gasoline from sunflower oil, which can be obtained from pyrolysis with 5% catalytic runs.

Figure 3.7 shows the curves of distillation of petroleum-based gasoline and gasoline from sunflower oil. The distillation curve of the gasoline from used lubricant oil by catalytic pyrolysis is similar to that of gasoline. Petroleum-based gasoline is slightly more volatile than sunflower oil gasoline.

3.7 Diesel-like Fuel from Tallow (Beef) by Pyrolysis and Steam Reforming

Tallow is a mixture of triglycerides, most of which are saturated; tristearin is usually the major component (Ma and Hanna, 1999). The tallow from animal sources is commonly used in soap production. In tallow the saturated fatty acid component accounts for almost 50% of the total fatty acids. The higher palmitic and stearic acid contents give tallow its unique properties of high melting point and high viscosity.

The first pyrolysis of vegetable oil was conducted in an attempt to synthesize petroleum from vegetable oil. Since World War I, many investigators have studied the pyrolysis of vegetable oils to obtain products suitable for fuel. In 1947, large-scale thermal cracking of tung oil calcium soaps was reported (Chang and Wan, 1947).

Used cottonseed oil from the cooking process was decomposed with Na_2CO_3 as catalyst at 725 K to give a pyrolyzate containing mainly C_{8-20} alkanes (69.6%) in addition to alkenes and aromatics. The pyrolyzate had lower viscosity, pour point, and flash point than D2 fuel and equivalent heating values (Bala, 2005).

Vegetable oil is converted into lower-molecular-weight products by two simultaneous reactions: cracking and condensation. The heavy hydrocarbons produced from primary and secondary deoxygenation and cracking are used to produce light olefins and light paraffins, water, carbon dioxide, and carbon monoxide. Hydrocarbon formation can be identified as deoxygenation, cracking, and aromatization with hydrogen transfer. Deoxygenation can take place via decarboxylation and dehydration (Change and Silvestri, 1977).

Certain petroleum fractions are converted into other kinds of chemical compounds. Catalytic isomerization converts straight-chain alkanes into branched-chain ones. The cracking process converts higher alkanes and alkenes, and thus increases the gasoline yield. The process of catalytic reforming converts alkanes and cycloalkanes into aromatic hydrocarbons and thus increases the gasoline yield. The process of catalytic reforming converts alkanes and cycloalkanes into aromatic hydrocarbons and thus provides the chief raw material for the large-scale synthesis of another broad class of compounds (Ramachandra, 2004).

Crude tallow (beef) samples were used in pyrolysis and steam-reforming procedures. The pyrolysis experiments were performed in a device designed for this purpose. The main element of this device was a tubular reactor of height 95.1 mm, ID 17.0 mm, and OD 19.0 mm inserted vertically into an electrically heated tubular furnace. Pyrolysis was carried out at heating rates of 10 K/s. For each run, the heater was started at ambient temperature and switched off when the desired temperature was reached. Pyrolysis runs were carried out on samples up to 5 g in a temperature range of 775 to 1025 K. The yields of pyrolysis products were determined gravimetrically by weighing the different fraction of char and oily products and of the gaseous fraction by a gas meter. The accuracy in the determination of the yields was about 3%. The temperature of the reaction vessel was measured with an iron-constantan thermocouple and controlled at ±3 K. The experiments were performed at 775, 775, 825, 875, 925, 975, and 1025 K. The steam reforming was performed in a 100-mL cylindrical autoclave made of 316 stainless steel. The sample was loaded from the bolt hole into the autoclave, and the hole was plugged with a screw bolt after each run. The experiments were carried out at 650, 700, 750, 800, and 850 K at various ratios of water to tallow: 0.5, 1.0, 1.5, and 2. At the end of every experiment, the heat was turned off and, once room temperature was reached, the carbonaceous residue (char) remaining and the condensed liquid in the collected apparatus of liquids could be weighed. The autoclave was supplied with heat from an external heater, and power was adjusted to give an approximate heating time of 60 min. The temperature of the reaction vessel was measured with an iron-constantan thermocouple and controlled at ±5 K. Reactions occurred during the heating period. After each run, the gas was vented and the autoclave was poured into a beaker. The rest of the oil and solids were removed from the autoclave by washing with used solvent. The products were recovered by washing with hot water. The obtained mixture was then filtered in a 30-mL glass crucible, with a medium frit, to separate the solvent and insoluble materials. The preweighed filter was dried to constant weight in an oven at 378 K.

Table 3.12 shows the average composition of fatty acids in tallow. The total fatty acids in the tallow samples were 51.1% by weight. The major fatty acids in the tallow were palmitic (28.7%), stearic (19.5), and oleic (44.4%). Higher palmitic and stearic acid contents give the tallow a high melting point.

Figure 3.8 shows the effect of temperature on total liquid and diesel-like liquid yields from tallow by pyrolysis. The reaction parameters of pyrolysis are tempera-

Table 3.12 Average composition of fatty acids in tallow (wt.%)

Fatty acid	%
Myristic (14:0)	2.9
Palmitic (16:0)	28.7
Stearic (18:0)	19.5
Oleic (18:1)	44.4
Linoleic (18:2)	3.6
Linolenic (18:3)	0.9

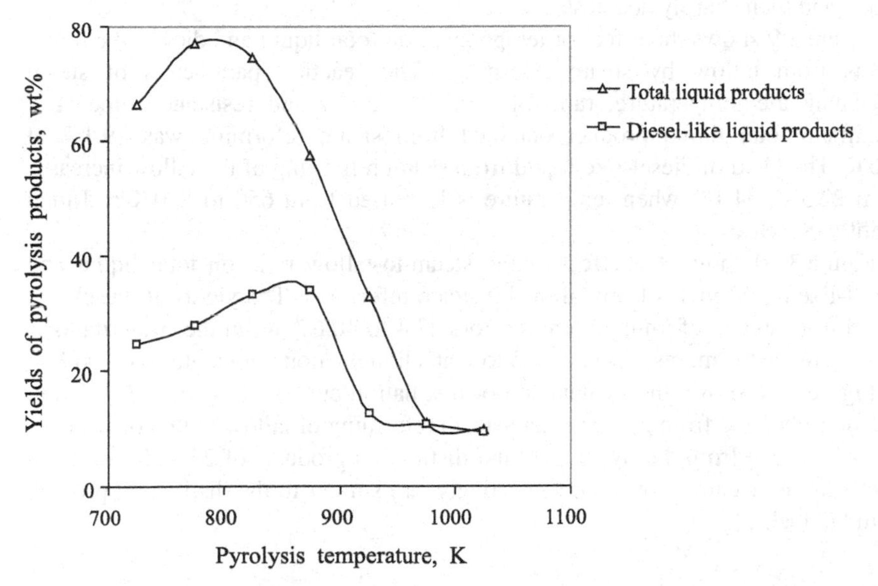

Fig. 3.8 Effect of temperature on total liquid and diesel-like liquid yields from tallow by pyrolysis (reaction time: 45 min)

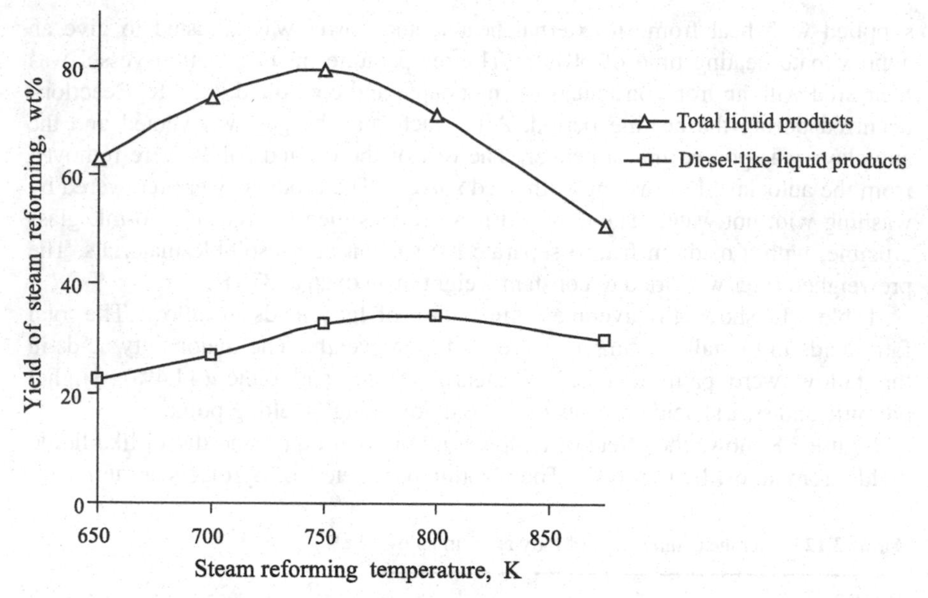

Fig. 3.9 Effect of temperature on total liquid and diesel-like liquid yields from tallow by steam reforming (reaction time: 45 min; steam-to-tallow ratio: 0.5)

ture and resistance time. The maximum total liquid product obtained from pyrolysis was 77.1% at 775 K. The yield of diesel-like liquid from pyrolysis of the tallow increases from 24.6 to 33.9% with increasing temperature from 725 to 875 K and then sharply decreases.

Figure 3.9 shows the effect of temperature on total liquid and diesel-like liquid yields from tallow by steam reforming. The reaction parameters of steam reforming are temperature, ratio of steam to tallow and resistance time. The maximum total liquid product obtained from steam reforming was 79.1% at 750 K. The yield of diesel-like liquid from steam reforming of the tallow increases from 22.3 to 34.1% when temperature is increased from 650 to 800 K and then slightly decreases.

Figure 3.10 shows the effect of the steam-to-tallow ratio on total liquid and diesel-like liquid yields from tallow by steam reforming. The yields of diesel-like liquid from steam reforming increase from 32.4 to 40.6% when the (water/tallow) ratio increases from, respectively, 0.5 to 2 at a liquefaction temperature of 800 K.

Figure 3.11 shows the comparison of distillation curves of average distillation products obtained from pyrolysis and steam reforming of tallow to that of D2 fuel. As can be seen from the figure, the first distillation products of 25 and 40% from pyrolysis and steam reforming, respectively, are similar to the distillation product from D2 fuel.

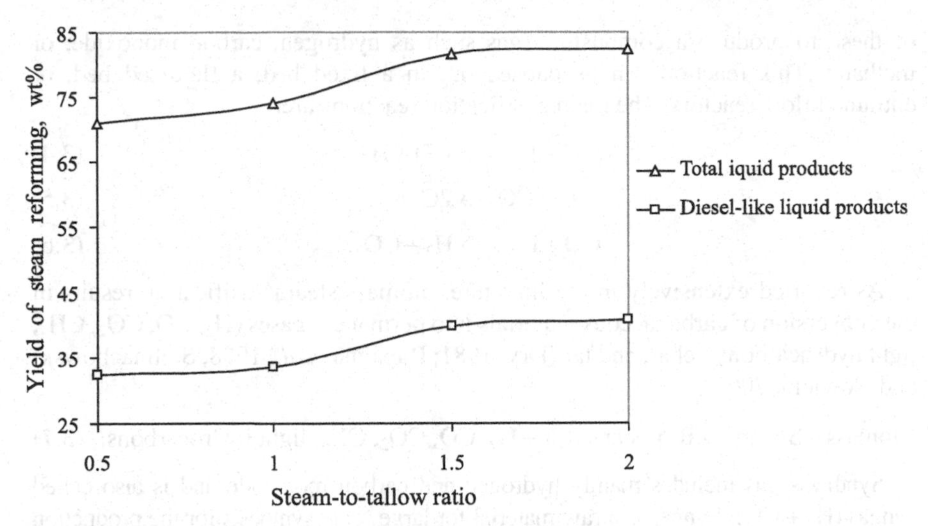

Fig. 3.10 Effect of steam-to-tallow ratio on total liquid and diesel-like liquid yields from tallow by supercritical water liquefaction (reaction time: 45 min; temperature: 800 K)

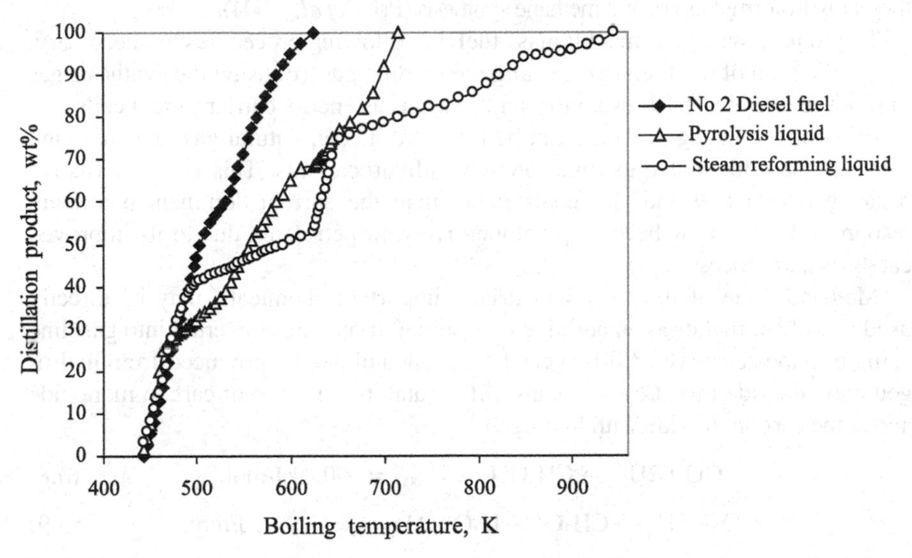

Fig. 3.11 Comparison of distillation curves of average distillation products obtained from pyrolysis and steam reforming of tallow to that of D2 fuel

3.8 Converting Triglyceride-derived Synthetic Gas to Fuels via Fischer–Tropsch Synthesis

Gasification is a thermochemical conversion of a biomass fuel to a gaseous fuel with a gasification agent such as air, oxygen, steam, carbon dioxide, or mixtures of these to produce a combustible gas such as hydrogen, carbon monoxide, or methane. This reaction can be carried out in a fixed bed, a fluidized bed, or entrained flow reactors. The main gasification reactions are:

$$C + H_2O \rightarrow CO + H_2, \tag{3.4}$$

$$C + CO_2 \rightarrow 2CO_2, \tag{3.5}$$

$$CO + H_2O \rightarrow H_2 + CO_2. \tag{3.6}$$

As reported extensively in the literature, biomass steam gasification results in the conversion of carbonaceous materials into permanent gases (H_2, CO, CO_2, CH_4, light hydrocarbons), char, and tar (Dry, 1981; Rapagna *et al.*, 1998; Stelmachowski and Nowicki, 2003).

$$\text{Biomass} + \text{Steam} \rightarrow \text{Bio-char} + \text{Tar} + H_2, CO, CO_2, CH_4, \text{light hydrocarbons}. \tag{3.7}$$

Synthesis gas includes mainly hydrogen and carbon monoxide and is also called syngas (H_2 + CO). Syngas is a raw material for large scale synthesis for the production of important base products of organic chemistry. The fundamental reactions of synthesis gas chemistry are methanol synthesis, Fischer–Tropsch synthesis (FTS), oxosynthesis (hydroformylation), and methane synthesis (Prins *et al.*, 2004).

To produce syngas from a biomass fuel the following procedures are necessary: (a) gasification of the fuel, (b) cleaning the product gas, (c) using the synthesis gas to produce chemicals, (d) using the synthesis gas as energy carrier in fuel cells.

FTS converts syngas, which can be made from coal, natural gas, biomass, and any carbonaceous materials, into long-chain hydrocarbons. This is an alternative route to obtain fuel and chemicals rather than the current dominant petroleum resources. FTS is now becoming competitive with petroleum due to its improved catalysts and processes.

Methanol, one of the most industrially important chemicals, may be directly used as a clean fuel or as an additive to gasoline. It may be converted into gasoline using a shape-selective (ZSM-5) catalyst. Methanol can be produced from hydrogen-carbon oxide mixtures by means of the catalytic reaction of carbon monoxide and some carbon dioxide with hydrogen:

$$CO + 2H_2 \rightarrow CH_3OH \quad \Delta H_{300\,K} = -90.8\,\text{kJ/mol}, \tag{3.8}$$

$$CO_2 + 3H_2 \rightarrow CH_3OH + H_2O \quad \Delta H_{300\,K} = -49.6\,\text{kJ/mol}. \tag{3.9}$$

The presence of a certain amount of CO_2 in the percentage range, 35 to 55 by volume, is necessary to optimize the reaction. Side reactions, also strongly exothermic, can lead to the formation of byproducts such as methane, dimethylether, or higher alcohols.

Methanol is currently produced on an industrial scale exclusively by catalytic conversion of synthetic gas ($H_2 + CO + CO_2$). Processes are classified according to the operating pressure: (a) low-pressure process (5 to 10 MPa), (b) medium–pressure process (10 to 25 MPa) and (c) high–pressure process (25 to 30 MPa). To produce one ton of methanol, 2.52×10^3 m^3 of synthesis gas (70% H_2, 21% CO,

7% CO_2) is necessary. CO hydrogenation or FTS on cobalt-based catalysts has been studied for over 70 years. Copper and zinc are the key components of methanol synthesis catalysts (Stelmachowski and Nowicki, 2003). Copper-zinc catalysts used in the low–pressure process require a sulfur-free gas ($H_2S < 1$ mL/m^3). The catalysts can be deactivated by sulfide ion (catalyst poison). Catalysts (ZnO and Cr_2O_3 activated with chromic acid) for the medium- and high-pressure processes can accept 30 mL/m^3 of H_2S.

Production of methanol with ZnO and Cr_2O_3 catalysts by the high-pressure process is no longer economical. Instead of these, copper-containing catalysts with higher activity and better selectivity is now used. Sulfur- and chlorine-containing pollution can be prevented by long use of copper-containing catalysts in industrial methanol production. These catalyst poisons must be removed from the feed gas mixture prior to methanol synthesis. The main advantages of the low-pressure process are lower investment and production costs, improved operational reliability, and greater flexibility in the choice of plant size.

The higher alcohol synthesizes from CO/H_2 mixtures for use as additives to gasoline to increase the octane number. Methanol and higher alcohols can be simultaneously produced from synthesis gas by using many different catalysts such as Cu/Zn, Zr/Fe, and Mo/Th. Similar experimental procedures can be used in FTS, HAS, and methanol synthesis. Three types of reaction are considered for HAS: formation of n-alcohols, formation of n-paraffins, and the water–gas shift.

Biomass can be converted into bio-syngas by non-catalytic, catalytic, and steam-gasification processes. FTS was established in 1923 by German scientists Franz Fischer and Hans Tropsch. The main aim of FTS is the synthesis of long-chain hydrocarbons from a CO and H_2 gas mixture. The FTS is described by the following set of equations (Anderson, 1984; Schulz, 1999; Sie and Krishna, 1999):

$$nCO + (n + m/2)\, H_2 \rightarrow C_nH_m + nH_2O, \tag{3.10}$$

where n is the average length of the hydrocarbon chain and m is the number of hydrogen atoms per carbon. All reactions are exothermic and the product is a mixture of different hydrocarbons in that paraffin and olefins are the main parts.

In FTS one mole of CO reacts with two moles of H_2 in the presence cobalt (Co)-based catalyst to afford a hydrocarbon chain extension ($-CH_2-$). The reaction of synthesis is exothermic ($\Delta H = -165$ kJ/mol):

$$CO + 2H_2 \rightarrow -CH_2- + H_2O \quad \Delta H = -165\ \text{kJ/mol} \tag{3.11}$$

The $-CH_2-$ is a building block for longer hydrocarbons. A main characteristic regarding the performance of FTS is the liquid selectivity of the process (Tijmensen *et al.*, 2002). For this reaction, given by Eq. (3.11), a H_2/CO ratio of at least 2 for the synthesis of the hydrocarbons is necessary. The reaction of the synthesis is exothermic ($\Delta H = -42$ kJ/mol).

Typical operation conditions for the FTS are a temperature range of 475 to 625 K and pressures of 15 to 40 bar, depending on the process. All reactions are exothermic. The kind and quantity of liquid product obtained is determined by the reaction temperature, pressure and residence time, type of reactor, and catalyst used. Iron catalysts have a higher tolerance for sulfur, are cheaper, and produce more olefin products and alcohols. However, the lifetime of the iron catalysts is short and in commercial installations generally limited to 8 weeks. Cobalt catalysts have the advantage of a higher conversion rate and a longer life (over 5 years). Cobalt catalysts are in general more reactive for hydrogenation and therefore produce fewer unsaturated hydrocarbons and alcohols compared to iron catalysts.

The products from FTS are mainly aliphatic straight-chain hydrocarbons (C_xH_y). In addition to C_xH_y, branched hydrocarbons, unsaturated hydrocarbons, and primary alcohols are also formed in minor quantities. The product distribution obtained from FTS includes the light hydrocarbons methane (CH_4), ethene (C_2H_4), and ethane (C_2H_6), LPG (C_3–C_4, propane and butane), gasoline (C_5–C_{12}), diesel fuel (C_{13}–C_{22}), and light waxes (C_{23}–C_{33}) (Table 3.1). Raw bio-syngas contains trace contaminants like NH_3, H_2S, HCl, dust, and alkalis in ash. The distribution of the products depends on the catalyst and the process parameters such as temperature, pressure, and residence time.

Figure 3.12 shows the yields of hydrogen and carbon monoxide obtained from the pyrolysis of tallow (beef) at different temperatures. Figure 3.12 shows how the yields of hydrogen and carbon monoxide from pyrolysis of tallow increases with increasing temperature. The yield of hydrogen from pyrolysis of the tallow sharply increases from 9.4 to 31.7% by volume of total gaseous products when the temperature is increased from 975 to 1175 K. The yield of carbon monoxide from pyrolysis increases from 20.6 to 26.7% by volume of total gaseous products when the temperature is increased from 1075 to 1175 K.

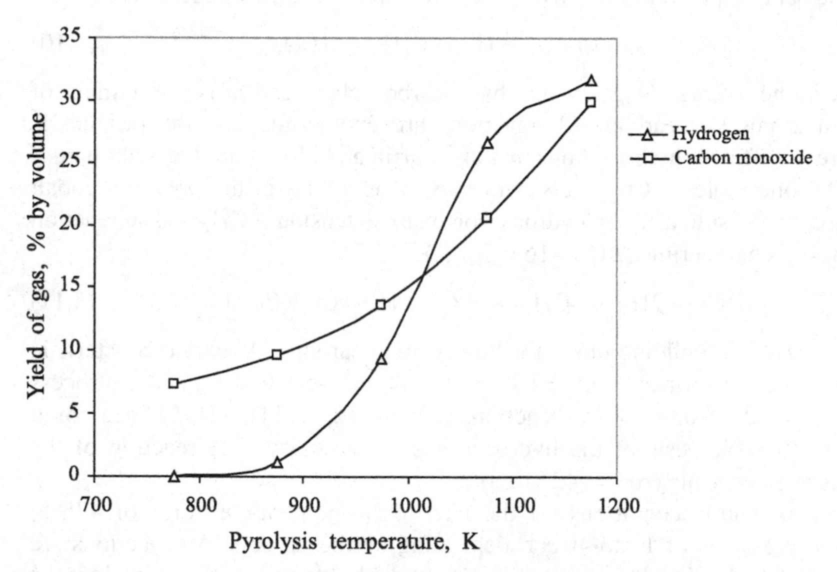

Fig. 3.12 Yields of hydrogen and carbon monoxide (% by volume of total gas products) obtained from pyrolysis of tallow (beef) at different temperatures

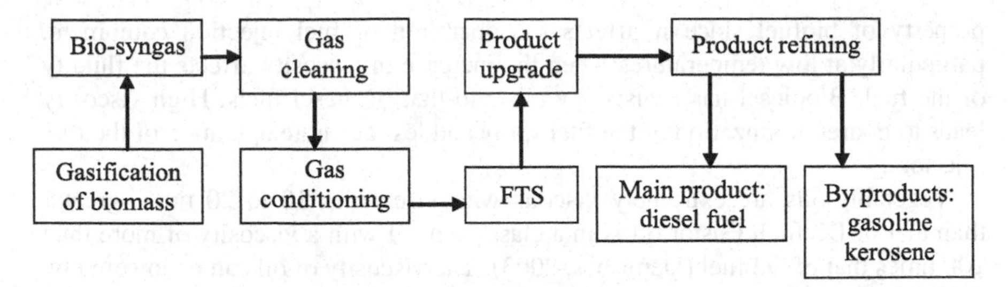

Fig. 3.13 Production of diesel fuel from bio-syngas by Fischer–Tropsch synthesis (FTS)

Hydrogen gas can be produced from the biomass material by direct and catalytic pyrolysis while, in one study, the final pyrolysis temperature was generally increased from 775 to 1025 K (Caglar, 2003). Hydrogen- and carbon-monoxide-rich gas products can be obtained from triglycerides by pyrolysis. The total yield of combustible gases (mainly H_2 and CO) for the triglyceride samples increased when the pyrolysis temperature was increased from 775 to 1175 K. The most important reaction parameters were temperature and resistance time.

Figure 3.13 shows the production of diesel fuel from bio-syngas by FTS. The design of a biomass gasifier integrated with a FTS reactor must be aimed at achieving a high yield of liquid hydrocarbons. For the gasifier, it is important to avoid methane formation as much as possible and convert all carbon in the biomass to mainly carbon monoxide and carbon dioxide (Prins *et al.*, 2004).

Gas cleaning is an important process before FTS. Gas cleaning is even more important for the integration of a biomass gasifier and a catalytic reactor. To avoid poisoning of FTS catalysts, tar, hydrogen sulfide, carbonyl sulfide, ammonia, hydrogen cyanide, alkali, and dust particles must be removed thoroughly (Tijmensen *et al.*, 2002).

3.9 Triglyceride Analyses

3.9.1 Viscosity

Viscosity is a measure of the internal fluid friction or resistance of oil to flow, which tends to oppose any dynamic change in the fluid motion (Song, 2000). As the temperature of oil increases, its viscosity decreases, and it is therefore able to flow more readily. It is also important for the flow of oil through pipelines, injector nozzles, and orifices (Radovanovic *et al.*, 2000). The lower the viscosity of the oil, the easier it is to pump and atomize and achieve finer droplets (Islam *et al.*, 2004).

Viscosity is measured on several different scales, including Redwood No. 1 at 100 F, Engler Degrees, Saybolt Seconds, *etc.* Viscosity is the most important property of biofuel since it affects the operation of fuel injection equipment, particularly at low temperatures when the increase in viscosity affects the fluidity of the fuel. Biodiesel has a viscosity close to that of diesel fuels. High viscosity leads to poorer atomization of the fuel spray and less accurate operation of the fuel injectors.

Vegetable oils are extremely viscous with viscosities 10 to 20 times greater than that of D2 fuel. Castor oil is in a class by itself with a viscosity of more than 100 times that of D2 fuel (Demirbas, 2003). The viscosity of oil can be lowered by blending it with pure ethanol. To reduce the high viscosity of vegetable oils, microemulsions with immiscible liquids such as methanol and ethanol and ionic or non-ionic amphiphiles have been studied (Ramadhas *et al.*, 2004; Mittelbach and Gangl, 2001). Short engine performances of both ionic and non-ionic microemulsions of ethanol in soybean oil were nearly as good as that of D2 fuel. All microemulsions with butanol, hexanol, and octanol met the maximum viscosity requirement for D2 fuel. 2-octanol was found to be an effective amphiphile in the micellar solubilization of methanol in triolein and soybean oil.

3.9.2 Density

Density is another important property of biofuel. Density is the mass per unit volume of any liquid at a given temperature. Specific gravity is the ratio of the density of a liquid to the density of water. Density has importance in diesel-engine performance since fuel injection operates on a volume metering system (Song, 2000). Also, the density of the liquid product is required for the estimation of the Cetane index (Srivastava and Prasad, 2000). In one study, densities were determined using a density meter at 298 K according to ASTM D5002-94. The density meter was calibrated using reverse osmosis water at room temperature.

3.9.3 Cetane Number

The cetane number (CN) is a measure of ignition quality or ignition delay and is related to the time required for a liquid fuel to ignite after injection into a compression ignition engine. CN is based on two compounds, namely, hexadecane, with a cetane of 100, and heptamethylnonane, with a cetane of 15. The CN scale also shows that straight-chain, saturated hydrocarbons have higher CNs than branched-chain or aromatic compounds of similar molecular weight and number of carbon atoms. The CN relates to the ignition delay time of a fuel upon injection into the combustion chamber. It is a measure of ignition quality of diesel fuels; a high CN implies short ignition delay. The longer the fatty acid carbon chains and the more saturated the molecules, the higher the CN. The CN of biofuel from animal fats is higher than those of vegetable oils. CN is determined from a real engine test. The cetane index (CI) is a calculated value derived from the density and volatility obtained from boiling characteristics of a fuel. CI usually gives a reasonably close approximation to a real CN (Song, 2000).

3.9.4 Cloud and Pour Points

Two important parameters for low-temperature applications of a fuel are cloud point (CP) and pour point (PP). The CP is the temperature at which a cloud of crystals first appears in a liquid when cooled under conditions as described in ASTM D2500-91. The PP is the temperature at which the amount of wax from solution is sufficient to gel the fuel; thus it is the lowest temperature at which the fuel can flow. The PP is the lowest temperature at which an oil specimen can still be moved. It is determined according to ASTM D97-96. These two properties are used to specify the cold-temperature usability of a fuel. In one study, two cooling baths with different cooling temperatures were used. Triglycerides have higher CP and PP compared to conventional diesel fuel (Prakash, 1998).

3.9.5 Distillation Range

The distillation range of a fuel affects its performance and safety. It is an important criterion for an engine's start and warmup. It is also needed in the estimation of the CI. The distillation range of the liquid product is determined by a test method (ASTM D2887-97) that covers the determination of the boiling range distribution of liquid fuels.

When the ASTM D86 procedure was used to distil vegetable oils, they cleaved to a two-phase distillate. Preliminary data indicate a complex mixture of products including alkanes, alkenes, and carboxylic compounds (Goering *et al.*, 1982). Typically, it is not possible to distil all of the vegetable oil, and some brownish residue remained in the distillation flask. However, soaps obtained from vegetable oils can be distilled into hydrocarbon-rich products with higher yields. The findings from distillation ranges of vegetable oils are given in the literature (Barsic and Humke, 1981).

3.9.6 Heat of Combustion

The heat of combustion measures the energy content in a fuel. This property is also referred to as calorific value or heating value. Although the CN determines combustion performance, it is the heating value, along with thermodynamic criteria, that sets the maximum possible output of power (Song, 2000). The higher heating values (HHVs) of oil samples are measured in a bomb calorimeter according to the ASTM D2015 standard method.

The ultimate analysis of a vegetable oil provides the weight percentages of carbon, hydrogen, and oxygen. The carbon, hydrogen, and oxygen contents of various common vegetable oils are 74.5 to 78.4, 10.6 to 12.4, and 10.8 to 12.0 wt.%, respectively. The HHV of vegetable oils (Goering *et al.*, 1982) ranges from 37.27 to 40.48 MJ/kg. The HHVs of various vegetable oils vary by <9%.

The saponification value (SV) of an oil decreases as its molecular weight increases. On the other hand, the percentages of carbon and hydrogen in oil increase as the molecular weight decreases. The increase in the iodine value (IV) (*i.e.*, carbon–carbon double bond, –C=C–, content) results in a decrease in the heat content of an oil. Therefore, for calculation of the HHVs (MJ/kg) of vegetable oils, Eq. (3.12) was suggested by Demirbas (1998):

$$HHV = 49.43 - [0.041(SV) + 0.015(IV)]. \quad (3.12)$$

3.9.7 Water Content

The water content of a fuel is required to accurately measure the net volume of actual fuel in sales, taxation, exchanges, and custody transfer (Srivastava and Prasad, 2000). Various methods are used for the determination of water content in oil samples such as evaporation methods, distillation methods, the xylene method, Karl-Fischer titration method, *etc.* Evaporation methods rely on measuring the mass of water in a known mass of sample. The moisture content is determined by measuring the mass of an oil sample before and after the water is removed by evaporation. Distillation methods are based on direct measurement of the amount of water removed from an oil sample by evaporation. The Karl-Fischer titration method is often used for determining the moisture content of oils that have low water content.

3.9.8 Discussion of Fuel Properties of Triglycerides

Fuel properties of triglyceride oils were characterized by determining its viscosity, density, cetane number, cloud and pour points, distillation range, flash point, ash content, sulfur content, carbon residue:, acid value, copper corrosion, and HHV (Goering *et al.*, 1982). Table 3.13 shows viscosity, density, flash point, and distillation range measurements of 15 vegetable oils. Figure 3.14 shows the relationships between viscosity and density for vegetable oils.

$$D = -0.7503V + 939.04, \quad (3.13)$$

Table 3.13 Viscosity, density, flash point, and distillation range measurements of 15 vegetable oils

Vegetable oil	Viscosity mm^2/s	Density g/L	Flash point K	Distillation range K
Ailanthus	30.2	916	513	423–623
Bay laurel	23.2	921	499	418–618
Beech	34.6	915	515	428–633
Beechnut	38.0	912	533	428–636
Corn	34.9	910	542	428–638
Cottonseed	33.5	915	524	443–638
Crambe	53.0	904	557	–
Hazelnut kernel	24.0	920	503	433–623
Linseed	27.2	924	520	438–638
Mustard oil	33.8	913	518	–
Peanut	39.6	903	543	440–644
Poppy seed	42.4	907	538	443–643
Rapeseed	37.3	912	531	–
Safflower seed	31.3	914	531	438–641
H.O. Safflower	41.2	902	548	–
Sesame	35.5	913	533	–
Soybean	32.6	914	528	–
Spruce	35.6	914	513	423–623
Sunflower seed	33.9	916	535	428–634
Walnut kernel	36.8	912	524	433–628

where D is the density and V the viscosity of an oil sample (Eq. 3.13). There is considerable regression between viscosity and density values of vegetable oils ($R^2 = 0.7494$).

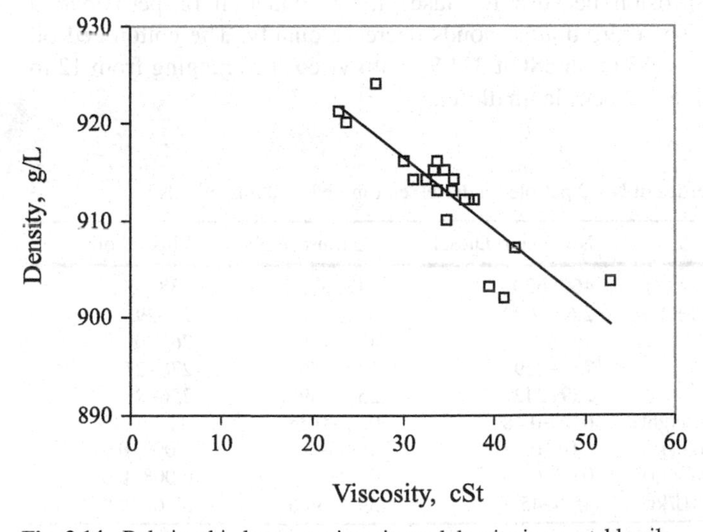

Fig. 3.14 Relationship between viscosity and density in vegetable oils

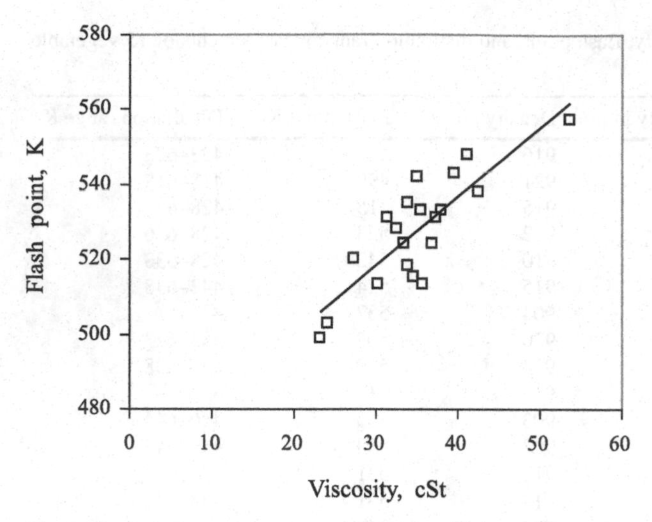

Fig. 3.15 Relationship between viscosity and flash point in vegetable oils

Figure 3.15 shows the relationship between viscosity and flash point in vegetable oils.

$$F = 1.8512V + 462.66, \quad (3.14)$$

where F is the flash point and V the viscosity of an oil sample (Eq. 3.14). There is regression between viscosity and flash point values of vegetable oils (R2 = 0.6996).

The fuel properties of No. 2 petroleum diesel, cottonseed, and linseed oils are given in Table 3.14. It is well known that a double bond in the fatty acid structure increases copper corrosion because it causes inner oxidation or peroxidation during combustion. Therefore double bonds decrease quality. The cottonseed oil was extremely viscous (33 to 36 cSt at 311 K) with viscosities ranging from 12 to 13 times greater than No. 2 petroleum diesel.

Table 3.14 Fuel properties of No. 2 petroleum diesel, cottonseed, and linseed oils

Property	No. 2 petr. Diesel	Cottonseed oil	Linseed oil
Distillation range (K)	460–620	443–628	438–638
Viscosity (mm^2/s at 311 K)	2.68–2.72	32–36	26–29
Cetane number	46–48	41–44.0	26–30
Cloud point (K)	257–259	274–276	276–279
Pour point (K)	239–242	257–259	256–259
Carbon residue (% by weight)	0.34–0.48	0.23–0.25	0.23–0.25
Ash content (% by weight)	0.01–0.02	0.008–0.01	0.009–0.01
Sulfur content (% by weight)	0.03–0.06	0.008–0.01	0.008–0.01
Higher heating value (MJ/kg)	45.0–45.3	39.4–39.6	39.6–39.8

The CN scale shows that straight-chain, saturated hydrocarbons have higher CNs compared to branched-chain or aromatic compounds of similar molecular weight and number of carbon atoms. The CN is one of the prime indicators of the quality of diesel fuel. It relates to the ignition delay time of a fuel upon injection into the combustion chamber. The CN is a measure of ignition quality of diesel fuels, and a high CN implies short ignition delay. In one study, the CN of CSO samples were in the range 41 to 44.0 (Table 3.14). The CN of biodiesel is generally higher than conventional diesel. The longer the fatty acid carbon chains and the more saturated the molecules, the higher the CN. The CN of biodiesel from animal fats is higher than those of vegetable oils (Bala, 2005).

3.10 Triglyceride Economy

High petroleum prices spur the study of biofuel production. Lower-cost feedstocks are needed since biodiesel from food-grade oils is not economically competitive with petroleum-based diesel fuel. Inedible plant oils have been found to be promising crude oils for the production of biodiesel.

The cost of biofuel and demand of vegetable oils can be reduced by inedible oils and used oils, instead of edible vegetable oil. Around the world large amounts of inedible oil plants are available in nature.

Vegetable oil is traditionally used as a natural raw material in linoleum, paint, lacquers, cosmetics, and laundry powder additives. There is a growing market in the field of lubricants, hydraulic oils, and special applications. Intensive use of pure plant oil in motors is an option to replace fossil fuels. Nowadays the technique is tested and well established. Pure plant oil fuel has the advantages of low sulfur and aromatic content and safer handling. The use of cold-pressed plant oil instead of fossil diesel leads to a reduction in the production of the greenhouse gas CO_2.

Everybody can produce their own fuel. The cold-pressing process does not require complicated machinery. The characteristics of this process are low energy requirements without any use of chemical extractive agents.

References

Adjaye, J.D., Katikaneni, S.P.R., Bakhsi, N.N. 1995. Catalytic conversion of canola oil to fuels and chemicals over various cracking catalysts. Can J Chem Eng 73:484–497.

Adjaye, J.D., Katikaneni, S.P.R., Bakhsi, N.N. 1996. Catalytic conversion of a biofuel to hydrocarbons: Effect of mixtures of HZSM-5 and silica-alumina catalysts on product distribution. Fuel Process Technol 48:115–143.

Agra, I.B., Warnijati, S., Pratama, M.S. 1992. Catalytic pyrolysis of nyamplung seeds oil to mineral oil like fuel. In: Sayigh, A.A.M. (ed.) Proceedings of the 2nd World Renewable Energy Congress. Pergamon, Reading, UK.

Alencar, J.W., Alves, P.B., Craveiro, A.A.1983. Pyrolysis of tropical vegetable oils. J Agric Food Chem 31:1268–1270.

Anderson, R.B. 1984. The Fischer-Tropsch synthesis. Academic, New York.
ASAE., 1982. Vegetable oil fuels. In: Backers, L. (ed.) Proceedings of the international conference on plant and vegetable oils as fuels. ASAE, St. Joseph, MI.
Association of Official Analytical Chemists (AOAC). 1993. Section C: Commercial Fats and Oils, American Oil Chemists Society (AOCS) official method Cd 1–25 for Iodine Value. Association of Official Analytical Chemists, Washington, D.C.
Association of Official Analytical Chemists (AOAC). 1997. Section C: Commercial Fats and Oils, American Oil Chemists Society (AOCS) official method Cd 3–25 for Saponification Value. Association of Official Analytical Chemists, Washington, D.C.
ASTM. 1979. Standard Test Method for Ignition Quality of Diesel Fuels by the Cetane Method, Designation: D 613, Annual Book of ASTM Standards, vol. 05.04, ASTM, Philadelphia, PA.
ASTM. 1988. Standard Test Method for Gross Calorific Value of Coal and Coke by the Adiabatic Bomb Calorimeter, Designation: D2015–85, Annual Book of ASTM Standards, vol. 05.02, ASTM, Philadelphia, PA, pp. 238–243.
ASTM. 1995a. Standard Test Method for Ash from Petroleum Products, Designation: D482-91, Annual Book of ASTM Standards, vol. 05.01, ASTM, Philadelphia, PA, pp. 198–199.
ASTM. 1995d. Standard Test Method for Determination of Total Sulfur in Light Hydrocarbons, Motor Fuels and Oils by Ultraviolet Fluorescence, Designation: D5453-93, Annual Book of ASTM Standards, vol. 05.03, ASTM, Philadelphia, PA, pp. 603–608.
ASTM D97-96 1998. Standard test method for pour point of petroleum products. In: Annual Book of ASTM Standards, vol. 05.01. ASTM, Philadelphia, PA, pp. 76–79.
ASTM D5002-94. 1998. Standard test method for calculated cetane index of distillate fuels. In: Annual Book of ASTM Standards, vol. 05.03. ASTM, Philadelphia, PA, pp. 263–266.
ASTM D2887-97. 1998. Standard test method for boiling range distribution of petroleum fractions by gas chromatography. In: Annual Book of ASTM Standards, vol. 05.03. ASTM, Philadelphia, PA, pp. 195–204.
Azam, M.M., Waris, A., Nahar, N.M. 2005. Prospects and potential of fatty acid methyl esters of some non-traditional seed oils for use as biodiesel in India. Biomass Bioenergy 29:293–302.
Bajpai, S., Prajapati, S., Luthra, R., Sharma, S., Naqvi, A., Kumar, S. 1999. Variation in the seed and oil yields and oil quality in the Indian germplasm of opium poppy Papaver somniferum. Genet Res Crop Evol 46:435–439.
Bala, B.K. 2005. Studies on biodiesels from transformation of vegetable oils for diesel engines. Energy Edu Sci Technol 15:1–43.
Barsic, N.J., Humke, A.L. 1981. Performance and emissions characteristics of a naturally aspirated diesel engine with vegetable oil fuels. SAE paper no. 810262. Society of Automotive Engineers, Warrendale, PA.
Barth, T. 1999. Similarities and differences in hydrous pyrolysis and source rocks. Org Geochem 30:1495-1507.
Bartholomew, D. 1981. Vegetable oil fuel. JAOCS 58:286A–288A.
Becker, E.W. 1994. In: Baddiley, J. *et al.* (eds.) Microalgae: biotechnology and microbiology. Cambridge University Press, Cambridge, UK.
Bhatia S., Twaiq F.A., Zabidi N.A.M. 1999. Catalytic conversion of palm oil to hydrocarbons: Performance of various zeolite catalysts. Ind Eng Chem Res 38:3230–3237.
Bhatia, S., Sang, O.Y., Twaiq, F., Zakaria, R., Mohamed, A.R. 2003. Biofuel production from catalytic cracking of palm oil. Energy Sour 25:859–869.
Bilgin, A., Durgun, O., Sahin Z. 2002. The effects of diesel-ethanol blends on diesel engine performance. Energy Sour 24:431–440.
Billaud, F., Dominguez, V., Broutin, P., Busson C. 1995. Production of hydrocarbons by pyrolysis of methyl esters from rapeseed oil. J Am Oil Chem Soc 72:1149.
Bridgwater, A.V., Meier, D., Radlein, D. 1999. An overview of fast pyrolysis of biomass. Org Geochem 30:1479–1493.
Brignole, E.A. 1986. Supercritical fluid extraction. Fluid Phase Equilibria 29:133–144.
Bringi, N.V. 1987. Non-traditional Oil Seed and Oils of India.. Oxford/IBH, New Delhi.
Caglar, A. 2003. Gaseous products from solid wastes. Energy Edu Sci Technol 10:107–110.

Calvin, M. 1985. Fuel oils from higher plants. Annu Proc Phytochem Soc Eur 26:147–160.
Chang, C.D., Silvestri, A.J. 1977. The conversion of methanol and other compounds to hydrocarbons over zeolite catalysts. J Catal 47:249–259.
Chang, C.C., Wan, S.W. 1947. China's motor fuels from tung oil. Ind Eng Chem 39:1543–1548.
Dandik, L., Aksoy, H.A. 1998. Pyrolysis of used sunflower oil in the presence of sodium carbonate by using fractionating pyrolysis reactor. Fuel Process Technol 57:81–92.
Darnoko, D., Cheryan, M. 2000. Kinetics of palm oil transesterification in a batch reactor. JAOCS 77:1263–1267.
Das, M., Das, S.K., Suthar, S.H. 2002. Composition of seed and characteristics of oil from Karingda [(Citrullus lanatus Thumb) Man of]. Int J Food Sci Technol 37:893–896.
Demirbas, A. 1991a. Analysis of beech wood fatty acids by supercritical acetone extraction. Wood Sci Technol 25:365–370.
Demirbas, A. 1991b. Fatty and resin acids recovered from spruce wood by supercritical acetone extraction. Holzforschung 45:337–339.
Demirbas, A. 1998. Fuel properties and calculation of higher heating values of vegetable oils. Fuel 77:1117–1120.
Demirbas, A. 2001. Mineral, protein, and fatty acids contents of hazelnut kernels. Energy Edu Sci Technol 7:37–43.
Demirbas, A. 2002a. Biodiesel from vegetable oils via transesterification in supercritical methanol. Energy Convers Mgmt 43:2349–56.
Demirbas, A. 2002b. Diesel fuel from vegetable oil via transesterification and soap pyrolysis. Energy Sour 24:835–841.
Demirbas, A. 2003a. Chemical and fuel properties of seventeen vegetable oils. Energy Sour 25:721–728.
Demirbas, A. 2003b. Biodiesel fuels from vegetable oils via catalytic and non-catalytic supercritical alcohol transesterifications and other methods: a survey. Energy Convers Mgmt 44:2093–2109.
Demirbas, A. 2003c. Fuel conversional aspects of palm oil and sunflower oil. Energy Sour 25:457–466.
Demirbas, A. 2004a. Recent advances in waste processing. Energy Edu Sci Technol 13:1–12.
Demirbas, A. 2004b. Pyrolysis of municipal plastic wastes for recovery of gasoline-range hydrocarbons. J Anal Appl Pyrol 72:97–102.
Demirbas, A. 2005. Pyrolysis of ground beech wood in irregular heating rate conditions. J Anal Appl Pyrol 73:39–43.
Demirbas, A. 2006. Biodiesel production via non-catalytic SCF method and biodiesel fuel characteristics. Energy Convers Mgmt 47:2271–2282.
Demirbas, A., Kara, H. 2006. New options for conversion of vegetable oils to alternative fuels. Energy Sour Part A Recov Util Environ Effects 28:619–626.
Dry, M.E. 1981. The Fischer-Tropsch Synthesis. In: Anderson, J.R., Boudart, M. (eds.) Catalysis-Science and Technology, Vol. 1, Springer, New York, p. 160.
Dykstra, G.J., Schwab, A.W., Selke, E., Sorenson, S.C., Pryde, E.H. 1988. Diesel fuel from thermal decomposition of soybean oil. J Am Oil Chem Soc 65:1781–1785.
Eisenmenger, M., Dunford, N., Eller, F., Taylor, S. 2005. Pilot scale supercritical carbon dioxide extraction and characterization of wheat germ oil. AOCS Proceedings, Salt Lake City, UT.
EPA (US Environmental Protection Agency), 2002. A comprehensive analysis of biodiesel impacts on exhaust emissions. Draft Technical Report, EPA420-P-02-001, October 2002.
Erickson, D.R., Pryde, E.H., Brekke, O.L., Mounts, T.L., Falb, R.A. 1980. Handbook of Soy Oil Processing and Utilization. American Soybean Association and the American Oil Chemists Society. St. Louis, Missouri and Champaign, IL.
Fang, T., Goto, M., Wang, X., Ding, X., Geng, J., Sasaki, M., Hirose, T. 2007. Separation of natural tocopherols from soybean oil byproduct with supercritical carbon dioxide. J Supercrit Fluids 40:50–58.
Foidl, N., Foidl, G., Sanchez, M., Mittelbach, M., Hackel, S. 1996. Jatropha curcas L. As a source for the production of biofuel in nicaragua. Bioresour Technol 58:77–82.

Fortes, I.C.P., Baugh, P.J. 1999. Study of analytical on-line pyrolysis of oils from macauba fruit (Acrocomia sclerocarpa M) via GC/MS. J Braz Chem Soc 10:469–477.
Giannelos, P.N., Zannikos, F., Stournas, S., Lois, E., Anastopoulos, G. 2002. Tobacco seed oil as an alternative diesel fuel: physical and chemical properties. Ind Crop Prod 16:1–9.
Goering, C.E. 1984. Final report for project on effect of nonpetroleum fuels on durability of direct- injection diesel engines under contract 59-2171-1-6-057-0, USDA, ARS, Peoria, IL.
Goering, E., Schwab, W.,. Daugherty, J., Pryde, H., Heakin, J. 1982. Fuel properties of eleven vegetable oils. Trans ASAE 25, 1472–1483.
Goodrich, J., Lawson, C., Lawson, V. P. 1980. Kashaya Pomo Plants. Heyday Books, Berkeley, CA .
Gubitz, G. M., Mittelbach, M., Trabi, M. 1999. Exploitation of the tropical seed plant Jatropha Curcas L. Bioresour Technol 67:73–82.
Gunstone, F.D., Hamilton, R.J. (eds.) 2001. Oleochemicals manufacture and applications. Sheffield Academic Press/CRC Press, Sheffield, UK/Boca Raton, FL.
Haas, M.J., Cichowicz, D.J., Dierov, J.K. 2001. Lipolytic activity of California-laurel (*Umbellularia californica*). J Am Oil Chem Soc 78:1067–1071.
He, H., Wang, T., Zhu, S. 2007. Continuous production of biodiesel fuel from vegetable oil using supercritical methanol process. Fuel 86:442–447.
Hoyer, G.C. 1985. Extraction with supercrtical fluids: Why, how, and so what? Chetech July:440–448
Islam, M. N., Islam, M.N., Beg, M.R.A. 2004. The fuel properties of pyrolysis liquid derived from urban solid wastes in Bangladesh. Bioresour Technol 92:181–186.
Karaosmonoglu, F. 1999. Vegetable oil fuels: a review. Energy Sour 21:221–231.
Karmee, S.K., Chadha, A. 2005. Preparation of biodiesel from crude oil of Pongamia pinnata. Bioresour Technol 96:1425–1429.
Katikaneni, S.P.R., Adjaye, J.D., Bakhshi, N.N. 1995. Catalytic conversion of canola oil to fuels and chemicals over various cracking catalysts. Can J Chem Eng 73:484–497.
Kekelidze, N.A. 1987. Essential oils of the bark and wood of the stems of Laurus nobilis. Chem Nat Compounds 23:384–385.
Kerschbaum, S., Rinke, G. 2004. Measurement of the temperature dependent viscosity of biodiesel fuels. Fuel 83:287–291.
Kilic, A., Altuntas, E. 2006. Wood and bark volatile compounds of Laurus nobilis L. Holz als Roh-und Werkstoff 64:317–320.
Knothe, G., Krahl, J., Van Gerpen, J. (eds.) 2005. The Biodiesel Handbook. AOCS Press, Champaign, IL.
Knothe, G., Sharp, C.A., Ryan, T.W. 2006. Exhaust emissions of biodiesel, petrodiesel, neat methyl esters, and alkanes in a new technology engine. Energy Fuels 20:403–408.
Komers, K., Stloukal, R., Machek, J., Skopal, F. 2001.Biodiesel from rapeseed oil, methanol and KOH 3. Analysis of composition of actual reaction mixture. Eur J Lipid Sci Technol 103:363–3471.
Leng, T.Y., Mohamed, A.R., Bhatia, S. 1999. Catalytic conversion of palm oil to fuels and chemicals. Can J Chem Eng 77:156–162.
Lima, D.G., Soares, V.C.D., Ribeiro, E.B., Carvalho, D.A., Cardoso, E.C.V., Rassi, F.C., Mundim, K.C., Rubim, J.C., Suarez, P.A.Z. 2004. Diesel-like fuel obtained by pyrolysis of vegetable oils. J Anal Appl Pyrol 71:987–996.
Ma, F., Hanna, M.A. 1999. Biodiesel production: a review. Bioresour Technol 70:1–15.
Machacon, H.T.C., Matsumoto, Y., Ohkawara, C., Shiga, S., Karasawa, T., Nakamura, H. 2001. The effect of coconut oil and diesel fuel blends on diesel engine performance and exhaust emissions. JSAE Rev 22:349–355.
Madras, G., Kolluru, C., Kumar, R. 2004. Synthesis of biodiesel in supercritical fluids. Fuel 83:2029–2033.
Maschio, G., Koufopanos, C., Lucchesi, A. 1992. Pyrolysis, a promising route for biomass utilization. Bioresour Technol 42:219–231.

Meher, L.C. Kulkarni, M.G. Dalai, A.K. Naik, S.N. 2006a. Transesterification of karanja (Pongamia pinnata) oil by solid basic catalysts. Eur J Lipid Sci Technol 108:3898–397.

Meher, L.C., Dharmagadda, V.S.S., Naik, S.N. 2006b. Optimization of alkali-catalyzed transesterification of Pongamia pinnata oil for production of biodiesel. Bioresour Technol 97:1392–1397.

Mohan, D., Pittman, Jr., C.U., Steele, P.H. 2006. Pyrolysis of wood/biomass for bio-oil: a critical review. Energy Fuels 20:848–889.

Mittelbach, M., Remschmidt, C. 2004. Biodiesels–The Comprehensive Handbook. Karl-Franzens University, Graz, Austria.

Mittelbach, M., Gangl, S. 2001. Long storage stability of biodiesel made from rapeseed and used frying oil. JAOCS 78:573–577.

Morrison, R.T., Boyd, R.N. 1983. Organic Chemistry, 4th edn. Allyn and Bacon, Singapore.

Nagai, K., Seko, T. 2000. Trends of motor fuel quality in Japan. JSAE Rev 21:457–462.

Nagel, N., Lemke, P. 1990. Production of methyl fuel from miceoalgea. Appl Biochem Biotechnol 24:355–361.

Niehaus, R.A., Goering, C.E., Savage, Jr., L.D., Sorenson, S.C. 1986. Cracked soybean oil as a fuel for a diesel engine. Trans ASAE 29:683–689.

Nitschke, W.R., Wilson, C.M. 1965. Rudolph Diesel, Pionier of the Age of Power. The University of Oklahoma Press, Norman, OK.

Paulaitis, M.E., Krukonis, V.J., Kurnik, R.T., Reid, R.C. 1983. Supercritical fluid extraction. Rev Chem Eng 1:181–248.

Penninger, J.M.L., Radosz, M., McHugh, M.A., Krukonis, V.J. (eds.) 1985. Supercritical fluid technology. Elsevier, Amsterdam.

Pioch, D., Lozano, P., Rasoanantoandro, M.C., Graille, J., Geneste, P., Guida, A. 1993. Biofuels from catalytic cracking of tropical vegetable oils. Oleagineux 48:289–291.

Prakash, C.B. 1998. A critical review of biodiesel as a transportation fuel in Canada. A Technical Report. GCSI – Global Change Strategies International, Canada.

Prins, M.J., Ptasinski, K.J., Janssen, F.J.J.G. 2004. Exergetic optimisation of a production process of Fischer–Tropsch fuels from biomass. Fuel Process Technol 86:375–389.

Pryde, E.H., 1983. Vegetable oil as diesel fuel: overview. JAOCS 60:1557–1558.

Pryor, R.W., Hanna, M.A., Schinstock, J.L., Bashford, L.L. 1982. Soybean oil fuel in a small diesel engine. Trans ASAE 26:333–338.

Radovanovic, M., Venderbosch, R.H., Prins, W., van Swaaij, W.P.M. 2000. Some remarks on the viscosity measurement of pyrolysis liquids. Biomass Bioenergy 18:209–222.

Rajaei, A., Barzegar, M., Yamini, Y. 2005. Supercritical fluid extraction of tea seed oil and its comparison with solvent extraction. Eur Food Res Technol 220:401–405.

Ramachandra, T.V. 2004. Conversion of vegetable oils to alternative diesel-like fuels. Energy Edu Sci Technol 14:33–42.

Ramadhas, A.S., Jayaraj, S., Muraleedharan, C. 2004. Biodiesel production from high FFA rubber seed oil. Fuel 84:335–340.

Ramadhas, A.S., Jayaraj, S., Muraleedharan, C. 2004. Use of vegetable oils as I.C. engine fuels—a review. Renew Energy 29:727–742. S. Rapagnà, N. Jand and P. U. Foscolo

Rapagna, S., Jand, N., Foscolo, P.U. 1998. Catalytic gasification of biomass to produce hydrogen rich gas. Int J Hydrogen Energy 23:551–557.

Reynolds, T., Dring, J.V., Hughes, C. 1991. Lauric acid-containing triglycerides in seeds of Umbellularia californica Nutt. (Lauraceae). J Am Oil Chem Soc 68:976–977.

Roselius, W., Vitzthum, O., Hubert, P. 1975. Methods of production cocoa butter. US Patent 3,923,847.

Sang, O.Y., Twaiq, F., Zakaria, R., Mohamed, A., Bhatia, S. 2003. Biofuel production from catalytic cracking of palm oil. Energy Sources 25:859–869.

Saucedo, E. 2001. Biodiesel. Ingeniera Quimica 20:19–29.

Schelenk, H., Gellerman, J.L. 1960. Esterification of fatty acids with diazomethane on a small scale. Anal Chem 32:1412–1414.

Schneider, G.M. 1978. Physicochemical principles of extraction with supercritical gases. Agnew Chem Int Ed Eng 17:716–727.
Schwab, A.W., Bagby, M.O., Freedman, B. 1987. Preparation and properties of diesel fuels from vegetable oils. Fuel 66:1372–1378.
Schwab, A.W., Dykstra, G.J., Selke, E., Sorenson, S.C., Pryde, E.H., 1988. Diesel fuel from thermal decomposition of soybean oil. JAOCS 65:1781–1786.
Sensoz, S., Angin, D., Yorgun, S. 2000. Influence of particle size on the pyrolysis of rapeseed (Brassica napus L.): fuel properties of bio-oil. Biomass Bioenergy 19:271–279.
Shay, E.G. 1993. Diesel fuel from vegetable oils: status and opportunities. Biomass Bioenergy 4:227–242.
Sheehan, J., Cambreco, V., Duffield, J., Garboski, M., Shapouri, H., 1998. An overview of biodiesel and petroleum diesel life cycles. A report by US Department of Agriculture and Energy, Washington, D.C., pp.1–35.
Schulz, H. 1999. Short history and present trends of FT synthesis. Appl Catal A General 186: 1–16
Sie, S.T., Krishna, R. 1999. Fundamentals and selection of advanced FT-reactors. Appl Catal A General 186:55–70.
Song, C. 2000. Introduction to chemistry of diesel fuels. In: Song, C., Hsu, C.S., Moshida, I. (eds.) Chemistry of Diesel Fuels. Taylor and Francis, London, p. 13.
Sonntag, N.O.V. 1979. Reactions of fats and fatty acids. In: Swern, D. (ed.) Bailey's Industrial Oil and Fat Products, Vol. 1, 4th edn. Wiley, New York, p. 99.
Srivastava and Prasad, R. 2000. Triglycerides-based diesel fuels. Renew Sust Energy Rev 4: 111–133
Stelmachowski, M., Nowicki, L. 2003. Fuel from the synthesis gas—the role of process engineering. Appl Energy 74:85–93.
Tijmensen, M.J.A., Faaij, A.P.C., Hamelinck, C.N., van Hardeveld, M.R.M. 2002. Exploration of the possibilities for production of Fischer Tropsch liquids and power via biomass gasification. Biomass Bioenergy 23:129–152.
Voelker, T.A., Worrell, A.C., Anderson, L., Bleibaum, J., Fan, C., Hawkins, D.J., Radke, S.E., Davies, H.M. 1992. Fatty acid biosynthesis redirected to medium chains in transgenic oilseed plants. Science 257:72–74.
Weisz, P.B., Haag, W.O., Rodeweld, P.G. 1979. Catalytic production of high-grade fuel (gasoline) from biomass compounds by shapedelective catalysis. Science 206:57–58.
Wikipedia. 2007. Neem oil characteristics.http://en.wikipedia.org/wiki/Neem_oil.
Wright, H.J., Segur, J.B., Clark, H.V., Coburn, S.K., Langdom, E.E., DuPuis, R.N. 1944. A report on ester interchange. Oil Soap 21:145–148.
Yarmo, M.A., Alimuniar, A., Ghani, R.A., Suliaman, A.R., Ghani, M., Omar, H., Malek, A. 1992. Transesterification products from the metathesis reaction of palm oil. J Mol Catal 76:373–379.
Ziejewski, M., Kaufman, K.R., Schwab, A.W., Pryde, E.H. 1984. Diesel engine evaluation of a nonionic sunflower oil-aqueous ethanol microemulsion. J Am Oil Chemists' Soc 61: 1620–1626.
Ziejewski, M., Goettler, H., Pratt, G.L. 1986. Paper No. 860301, International Congress and Exposition, Detroit, MI, 24–28 February 1986.
Zhenyi, C., Xing, J., Shuyuan, L., Li, L. 2004. Thermodynamics calculation of the pyrolysis of vegetable oils. Energy Sour 26:849–856.

Chapter 4
Biodiesel

4.1 Introduction to Biodiesel Concept

The scarcity of conventional fossil fuels, growing emissions of combustion-generated pollutants, and their increasing costs will make biomass sources more attractive (Sensoz *et al.*, 2000). On the other hand, biomass use, in which many people already have an interest, has the properties of being a biomass source and a carbon-neutral source (Dowaki *et al.*, 2007). Experts suggest that current oil and gas reserves would suffice to last only a few more decades. To meet the rising energy demand and replace reducing petroleum reserves, fuels such as biodiesel and bioethanol are in the forefront of alternative technologies. Accordingly, the viable alternative for compression-ignition engines is biodiesel.

Biodiesel is briefly defined as the monoalkyl esters of vegetable oils or animal fats. Biodiesel is the best candidate for diesel fuels in diesel engines. Biodiesel burns like petroleum diesel as it involves regulated pollutants. On the other hand biodiesel probably has better efficiency than gasoline. Biodiesel also exhibits great potential for compression-ignition engines. Diesel fuel can also be replaced by biodiesel made from vegetable oils. Biodiesel is now mainly being produced from soybean, rapeseed, and palm oils. The higher heating values (HHVs) of biodiesels are relatively high. The HHVs of biodiesels (39 to 41 MJ/kg) are slightly lower than those of gasoline (46 MJ/kg), petrodiesel (43 MJ/kg), or petroleum (42 MJ/kg), but higher than coal (32 to 37 MJ/kg).

Biodiesel is pure, or 100%, biodiesel fuel. It is referred to as B100 or "neat" fuel. A biodiesel blend is pure biodiesel blended with petrodiesel. Biodiesel blends are referred to as BXX. The XX indicates the amount of biodiesel in the blend (*i.e.*, a B80 blend is 80% biodiesel and 20% petrodiesel).

The majority of energy demand is fulfilled by conventional energy sources like coal, petroleum, and natural gas. Petroleum-based fuels are limited reserves concentrated in certain regions of the world. These sources are on the verge of reaching their peak production. The scarcity of known petroleum reserves will make renewable energy sources more attractive (Sheehan *et al.*, 1998).

World energy demand continues to rise. The most feasible way to meet this growing demand is by using alternative fuels. One such fuel that exhibits great potential is biofuel, in particular biodiesel (Fernando *et al.*, 2006). The term biofuel can refer to liquid or gaseous fuels for the transport sector that are predominantly produced from biomass (Demirbas, 2006). Biofuels include energy security reasons, environmental concerns, foreign exchange savings, and socioeconomic issues related to the rural sector (Reijnders, 2006). In developed countries there is a growing trend toward using modern technologies and efficient bioenergy conversion using a range of biofuels, which are becoming costwise competitive with fossil fuels (Puhan *et al.*, 2005).

It is well known that transport is almost totally dependent on fossil-, particularly petroleum-, based fuels such as gasoline, diesel fuel, liquefied petroleum gas (LPG), and natural gas (NG). An alternative fuel to petrodiesel must be technically feasible, economically competitive, environmentally acceptable, and easily available. The current alternative diesel fuel can be termed biodiesel. Biodiesel use may improve emissions levels of some pollutants and deteriorate others. However, for quantifying the effect of biodiesel it is important to take into account several other factors such as raw material, driving cycle, vehicle technology, *etc.* Use of biodiesel will allow a balance to be sought between agriculture, economic development, and the environment (Meher *et al.*, 2006).

4.2 History

The process for making fuel from biomass feedstock used in the 1800s is basically the same one used today. The history of biodiesel is more political and economical than technological. The early 20th century saw the introduction of gasoline-powered automobiles. Oil companies were obliged to refine so much crude oil to supply gasoline that they were left with a surplus of distillate, which is an excellent fuel for diesel engines and much less expensive than vegetable oils. On the other hand, resource depletion has always been a concern with regard to petroleum, and farmers have always sought new markets for their products. Consequently, work has continued on the use of vegetable oils as fuel.

Producing biodiesel from vegetable oils is not a new process. The conversion of vegetable oils or animal fats into monoalkyl esters or biodiesel is known as transesterification. Transesterification of triglycerides in oils is not a new process. Duffy and Patrick conducted Transesterification as early as 1853. Life for the diesel engine began in 1893, when the famous German inventor Dr. Rudolph Diesel published a paper entitled “The theory and construction of a rational heat engine”. The paper described a revolutionary engine in which air would be compressed by a piston to a very high pressure, thereby causing a high temperature. Dr. Diesel designed the original diesel engine to run on vegetable oil.

Dr. Diesel was educated at the predecessor school to the Technical University of Munich in Germany. In 1878, he was introduced to the work of Sadi Carnot, who theorized that an engine could achieve much higher efficiency than the steam engines of the day. Diesel sought to apply Carnot's theory to the internal combustion engine. The efficiency of the Carnot cycle increases with the compression ratio—the ratio of gas volume at full expansion to its volume at full compression. Nicklaus Otto invented an internal combustion engine in 1876 that was the predecessor to the modern gasoline engine. Otto's engine mixed fuel and air before their introduction to the cylinder, and a flame or spark was used to ignite the fuel-air mixture at the appropriate time. However, air gets hotter as it is compressed, and if the compression ratio is too high, the heat of compression will ignite the fuel prematurely. The low compression ratios needed to prevent premature ignition of the fuel-air mixture limited the efficiency of the Otto engine. Dr. Diesel wanted to build an engine with the highest possible compression ratio. He introduced fuel only when combustion was desired and allowed the fuel to ignite on its own in the hot compressed air. Diesel's engine achieved efficiency higher than that of the Otto engine and much higher than that of the steam engine. Diesel received a patent in 1893 and demonstrated a workable engine in 1897. Today, diesel engines are classified as "compression-ignition" engines, and Otto engines are classified as "spark-ignition" engines.

Dr. Diesel used peanut oil to fuel one of his engines at the Paris Exposition of 1900 (Nitschke and Wilson, 1965). Because of the high temperatures created, the engine was able to run a variety of vegetable oils including hemp and peanut oil. At the 1911 World's Fair in Paris, Dr. Diesel ran his engine on peanut oil and declared "the diesel engine can be fed with vegetable oils and will help considerably in the development of the agriculture of the countries which use it". One of the first uses of transesterified vegetable oil was powering heavy-duty vehicles in South Africa before World War II. The name "biodiesel" has been given to transesterified vegetable oil to describe its use as a diesel fuel (Demirbas, 2002). Vegetable oils were used in diesel engines until the 1920s. During the 1920s, diesel engine manufacturers altered their engines to utilize the lower viscosity of petrodiesel, rather than vegetable oil.

The use of vegetable oils as an alternative renewable fuel competing with petroleum was proposed in the early 1980s. The advantages of vegetable oils as diesel fuel are its portability, ready availability, renewability, higher heat content (about 88% of No. 2 petroleum diesel fuel), lower sulfur content, lower aromatic content, and biodegradability. The energy supply concerns of the 1970s renewed interest in biodiesel, but commercial production did not begin until the late 1990s.

Dr. Diesel believed that engines running on plant oils had potential and that these oils could one day be as important as petroleum-based fuels. Since the 1980s, biodiesel plants have opened in many European countries, and some cities have run buses on biodiesel, or blends of petro and biodiesels. More recently, Renault and Peugeot have approved the use of biodiesel in some of their truck engines. Recent environmental and domestic economic concerns have prompted a resurgence in the use of biodiesel throughout the world. In 1991, the European Community (EC) proposed a 90% tax deduction for the use of biofuels, including biodiesel. Biodiesel plants are now being built by several companies in Europe;

each of these plants will produce up to 1.5 million gallons of fuel per year. The European Union accounted for nearly 89% of all biodiesel production worldwide in 2005.

4.3 Definitions

Biodiesel (Greek, bio, life + diesel from Rudolf Diesel) refers to a diesel-equivalent, processed fuel derived from biological sources. Biodiesel is the name for a variety of ester-based oxygenated fuels from renewable biological sources. It can be made from processed organic oils and fats.

Chemically, biodiesel is defined as the monoalkyl esters of long-chain fatty acids derived from renewable biolipids. Biodiesel is typically produced through the reaction of a vegetable oil or animal fat with methanol or ethanol in the presence of a catalyst to yield methyl or ethyl esters (biodiesel) and glycerine (Demirbas, 2002). Fatty acid (m)ethyl esters or biodiesels are produced from natural oils and fats. Generally, methanol is preferred for transesterification because it is less expensive than ethanol (Graboski and McCormick, 1998).

In general terms, biodiesel may be defined as a domestic, renewable fuel for diesel engines derived from natural oils like soybean oil that meets the specifications of ASTM D 6751. In technical terms (ASTM D 6751) biodiesel is a diesel engine fuel comprised of monoalkyl esters of long-chain fatty acids derived from vegetable oils or animal fats, designated B100 and meeting the requirements of ASTM D 6751. Biodiesel, in application as an extender for combustion in CIEs (diesel), possesses a number of promising characteristics, including reduction of exhaust emissions (Dunn, 2001). Chemically, biodiesel is referred to as the monoalkyl esters, especially (m)ethylester, of long-chain fatty acids derived from renewable lipid sources via a transesterification process.

Table 4.1 Technical properties of biodiesel

Common name	Biodiesel (bio-diesel)
Common chemical name	Fatty acid (m)ethyl ester
Chemical formula range	C_{14}–C_{24} methyl esters or $C_{15-25}H_{28-48}O_2$
Kinematic viscosity range (mm^2/s, at 313 K)	3.3–5.2
Density range (kg/m^3, at 288 K)	860–894
Boiling point range (K)	>475
Flash point range (K)	430–455
Distillation range (K)	470–600
Vapor pressure (mm Hg, at 295 K)	<5
Solubility in water	Insoluble in water
Physical appearance	Light to dark yellow, clear liquid
Odor	Light musty/soapy odor
Biodegradability	More biodegradable than petroleum diesel
Reactivity	Stable, but avoid strong oxidizing agents

Biodiesel is a mixture of methyl esters of long-chain fatty acids like lauric, palmitic, stearic, oleic, *etc.* Typical examples are rapeseed oil, canola oil, soybean oil, sunflower oil, palm oil, and their derivatives from vegetable sources. Beef and sheep tallow and poultry oil from animal sources and cooking oil are also sources of raw materials. The chemistry of conversion into biodiesel is essentially the same. Oil or fat reacts with methanol or ethanol in the presence of a sodium hydroxide or potassium hydroxide catalyst to form biodiesel, (m)ethylesters, and glycerine.

4.4 Biodiesel as an Alternative to Diesel Engine Fuel

Biodiesel is a processed fuel that can be readily used in diesel-engine vehicles, which distinguishes biodiesel from the straight vegetable oils or waste vegetable oils used as fuels in some modified diesel vehicles. In general, the physical and chemical properties and the performance of ethyl esters are comparable to those of the methyl esters. Methyl and ethyl esters have almost the same heat content. The viscosities of ethyl esters are slightly higher and the cloud and pour points are slightly lower than those of the methyl esters. Engine tests have demonstrated that methyl esters produce slightly higher power and torque than ethyl esters.

Biodiesel is a clear amber-yellow liquid with a viscosity similar to that of petrodiesel. Biodiesel is non-flammable and, in contrast to petrodiesel, is non-explosive, with a flash point of 423 K for biodiesel as compared to 337 K for petrodiesel. Unlike petrodiesel, biodiesel is biodegradable and non-toxic, and it significantly reduces toxic and other emissions when burned as a fuel. Currently, biodiesel is more expensive to produce than petrodiesel, which appears to be the primary factor in preventing its more widespread use. Current worldwide production of vegetable oil and animal fat is not enough to replace liquid fossil fuel use (maximum replacement percentage: *ca.* 20 to 25%) (Bala, 2005).

Methyl esters of vegetable oils (biodiesels) have several outstanding advantages among other new-renewable and clean-engine fuel alternatives. Methanol as a monoalcohol is generally used in the transesterification reaction of triglycerides in the presence of alkali as a catalyst (Clark *et al.*, 1984). Methanol is a relatively inexpensive alcohol. Several common vegetable oils such as sunflower, palm, rapeseed, soybean, cottonseed, and corn oils and their fatty acids can be used as the sample vegetable oil. Biodiesel is easier to produce and cleaner with equivalent amounts of processing when starting with clean vegetable oil. Tallow, lard, and yellow grease biodiesels require additional processing at the end of the transesterification process due to the presence of high free fatty acids. Diesel derived from rapeseed oil is the most common biodiesel available in Europe, while soybean biodiesel predominates in the United States.

The emergence of transesterification can be dated back to as early as 1846 when Rochieder described glycerol preparation through the ethanolysis of castor oil (Formo, 1979). Since that time alcoholysis has been studied in many parts of the world. Other researchers have also investigated the important reaction conditions and parameters in the alcoholysis of triglycerides such as fish oils, tallow, soybean, rapeseed, cottonseed, sunflower, safflower, peanut, and linseed oils (Fuls and Hugo, 1981; Freedman and Pryde, 1982; Isigigur *et al.*, 1994; Lang *et al.*, 2001; Mittelbach and Gangl, 2001; Goodrum, 2002).

The advantages of biodiesel as diesel fuel are its portability, ready availability, renewability, higher combustion efficiency, and lower sulfur and aromatic content (Ma and Hanna, 1999; Knothe *et al.*, 2006), higher cetane number, and higher biodegradability (Mudge and Pereira, 1999; Speidal *et al.*, 2000; Zhang *et al.*, 2003). The main advantages of biodiesel given in the literature include its domestic origin, which would help reduce a country's dependency on imported petroleum, its biodegradability, high flash point, and inherent lubricity in the neat form (Mittelbach and Remschmidt, 2004; Knothe *et al.*, 2005).

The major disadvantages of biodiesel are its higher viscosity, lower energy content, higher cloud point and pour point, higher nitrogen oxide (NO_x) emissions, lower engine speed and power, injector coking, engine compatibility, high price, and greater engine wear. The technical disadvantages of biodiesel/fossil diesel blends include problems with fuel freezing in cold weather, reduced energy density, and degradation of fuel under storage for prolonged periods. One additional problem is encountered when blends are first introduced into equipment that has a long history of pure hydrocarbon usage. Hydrocarbon fuels typically form a layer of deposits on the inside of tanks, hoses, *etc*. Biodiesel blends loosen these deposits, causing them to block fuel filters. However, this is a minor problem, easily remedied by proper filter maintenance during the period following introduction of the biodiesel blend (Wardle, 2003).

Direct use of vegetable oils and/or the use of blends of the oils has generally been considered to be not satisfactory and impractical for both direct and indirect diesel engines. The high viscosity, acid composition, free fatty acid content, and gum formation due to oxidation and polymerization during storage and combustion, carbon deposition, and oil thickening are obvious problems (Ma and Hana, 1999).

Biodiesel has significant potential for use as an alternative fuel in compression-ignition engines (Demirbas, 2003; Knothe *et al.*, 1997). Biofuels are non-toxic, biodegradable, and free of sulfur and carcinogenic compounds (Venkataraman, 2002) as they are obtained from renewable sources. Biodiesel is a plant-derived product and contains oxygen in its molecules, making it a cleaner-burning fuel than petrol and diesel (Sastry *et al.*, 2006).

Biodiesel is a clean-burning alternative fuel produced from domestic, renewable resources that are much more efficient to produce and use than gasoline. The developmental history of biodiesel is more political than technological. The actual process for making biodiesel was originally developed in the early 1800s and has basically remained unchanged. It was the political and economic influences of industrial leaders during the 1920s and 1930s that caused the fuel trends to favor the use of petroleum-based fuels as opposed to agricultural fuels.

Table 4.2 shows the availability of modern transportation fuels. The advantage of biodiesel in this aspect is that it is a derivative of natural products. As demand rises, the production of the required agricultural products can be increased to

Table 4.2 Availability of modern transportation fuels

Fuel type	Availability	
	Current	Future
Gasoline	Excellent	Moderate-poor
Biodiesel	Moderate	Excellent
Compressed natural gas (CNG)	Excellent	Moderate
Hydrogen fuel cell	Poor	Excellent

compensate. Biodiesel is a technologically feasible alternative to fossil diesel, but nowadays biodiesel costs 1.5 to 3 times more than fossil diesel. As far as actual fuel costs are concerned, the cost of biodiesel currently is comparable to that of gasoline. Biodiesel will be a reasonably available engine fuel in the near future.

Biodiesel has got better lubricant properties than fossil diesel. Its oxygen content improves the combustion process, leading to a decreased level of tailpipe polluting emissions. Biodiesel is non-toxic and quickly biodegrades. The risks of handling, transporting, and storing biodiesel are much lower than those associated with fossil diesel. The competitiveness of biodiesel relies on the prices of biomass feedstock and costs, which are linked to conversion technology. Depending on the feedstock used, byproducts may have more or less relative importance. Biodiesel is not competitive with fossil diesel under current economic conditions, where the positive externalities, such as impacts on the environment, employment, climate changes, and trade balance, are not reflected in the price mechanism.

Biodiesel from virgin vegetable oil reduces carbon dioxide emissions and petroleum consumption when used in place of petroleum diesel. This conclusion is based on a life cycle analysis of biodiesel and petroleum diesel, accounting for resource consumption and emissions for all steps in the production and use of the fuel.

4.5 Sources of Biodiesel

There are various other biodiesel sources: almond, andiroba (*Carapa guianensis*), babassu (*Orbignia sp.*), barley, camelina (*Camelina sativa*), coconut, copra, cumaru (*Dipteryx odorata*), *Cynara cardunculus*, fish oil, groundnut, *Jatropha curcas*, karanja (*Pongamia glabra*), laurel, *Lesquerella fendleri*, *Madhuca indica*, microalgae (*Chlorella vulgaris*), oat, piqui (*Caryocar sp.*), poppy seed, rice, rubber seed, sesame, sorghum, tobacco seed, and wheat (Pinto *et al.*, 2005).

Vegetable oils are renewable fuels. They have become more attractive recently because of their environmental benefits and the fact that they are made from renewable resources. Vegetable oils are a renewable and potentially inexhaustible source of energy, with energy content close to that of diesel fuel. Global vegetable oil production increased from 56 million tons in 1990 to 88 million tons in 2000, following a below-normal increase. The source of this gain was distributed among the various oils. Global consumption rose 56 million tons to 86 million tons, leaving world stocks comparatively tight.

A variety of biolipids can be used to produce biodiesel. These are (a) virgin vegetable oil feedstock; rapeseed and soybean oils are most commonly used, though other crops such as mustard, palm oil, sunflower, hemp, and even algae show promise; (b) waste vegetable oil; (c) animal fats including tallow, lard, and yellow grease; and (d) non-edible oils such as jatropha, neem oil, castor oil, tall oil, *etc.*

The widespread use of soybeans in the USA for food products has led to the emergence of soybean biodiesel as the primary source for biodiesel in that country. In Malaysia and Indonesia, palm oil is used as a significant biodiesel source. In Europe, rapeseed is the most common base oil used in biodiesel production. In India and southeast Asia, the jatropha tree is used as a significant fuel source.

Algae can grow practically anywhere where there is enough sunshine. Some algae can grow in saline water. The most significant distinguishing characteristic of algal oil is its yield and hence its biodiesel yield. According to some estimates, the yield (per acre) of oil from algae is over 200 times the yield from the best-performing plant/vegetable oils (Sheehan *et al.*, 1998). Microalgae are the fastest-growing photosynthesizing organisms. They can complete an entire growing cycle every few days. Approximately 46 tons of oil/hectare/year can be produced from diatom algae. Different algae species produce different amounts of oil. Some algae produce up to 50% oil by weight. The production of algae to harvest oil for biodiesel has not been undertaken on a commercial scale, but working feasibility studies have been conducted to arrive at the above number.

Specially bred mustard varieties can produce reasonably high oil yields and have the added benefit that the meal left over after the oil has been pressed out can act as an effective and biodegradable pesticide.

References

Bala, B.K. 2005. Studies on biodiesels from transformation of vegetable oils for diesel engines. Edu Sci Technol 15:1–43.

Clark, S.J., Wagner, L., Schrock, M.D., Pinnaar, P.G. 1984. Methyl and ethyl esters as renewable fuels for diesel engines. JAOCS 61:1632–1638.

Demirbas, A. 2002. Biodiesel from vegetable oils via transesterification in supercritical methanol. Convers Mgmt 43:2349–56.

Demirbas, A. 2003. Biodiesel fuels from vegetable oils via catalytic and non-catalytic supercritical alcohol transesterifications and other methods: a survey. Energy Convers Mgmt 44:2093–2109.

Demirbas, A. 2006. Global biofuel strategies. Energy Edu Sci Technol 17:27–63.

Dowaki, K., Ohta, T., Kasahara, Y., Kameyama, M., Sakawaki, K., Mori, S. 2007. An economic and energy analysis on bio-hydrogen fuel using a gasification process. Renewable Energy 32:80–94

Dunn, R. O. 2001. Alternative jet fuels from vegetable-oils. Trans ASAE 44:1151-757.

Fernando, S., Hall, C., Jha, S., 2006. NO_x reduction from biodiesel fuels. Energy Fuels 20: 376–382

Formo, M.W. 1979. Physical properties of fats and fatty acids. Bailey's Industrial Oil and Fat Products. Vol.1, 4th edn. Wiley, New York.
Freedman, B., Pryde, E.H. 1982. Fatty esters from vegetable oils for use as a diesel fuel. In: Proceedings of the international conference on plant and vegetable oils as fuels, pp. 17–122.
Fuls, J., Hugo, F.J.C. 1981. On farm preparation of sunflower oil esters for fuel. In: Proceedings of the 3rd international conference on energy use management, pp. 1595–1602.
Goodrum, J.W. 2002. Volatility and boiling points of biodiesel from from vegetable oils and tallow. Bioenergy 22:205–211.
Graboski, M.S., McCormick, R.L. 1998. Combustion of fat and vegetable oil derived fuels in diesel engines. Prog Energy Combust Sci 24:125–164.
Isigigur, A., Karaosmonoglu, F., Aksoy, H. A. 1994. Methyl ester from safflower seed oil of Turkish origin as a biofuel for diesel engines. Appl Biochem Biotechnol 45/46:103–112.
Knothe, G., Dunn, R.O., Bagby, M.O. 1997. Biodiesel: the use of vegetable oils and their derivatives as alternative diesel fuels. Am Chem Soc Symp Ser 666:172–208.
Knothe, G., Krahl, J., Van Gerpen, J. (eds.) 2005. The Biodiesel Handbook. AOCS Press: Champaign, IL.
Knothe, G., Sharp, C.A., Ryan, T.W. 2006. Exhaust emissions of biodiesel, petrodiesel, neat methyl esters, and alkanes in a new technology engine. Energy Fuels 20:403–408.
Lang, X., Dalai, A.K., Bakhshi, N.N., Reaney, M.J., Hertz, P.B. 2001. Preparation and characterization of bio-diesels from various bio-oils. Bioresour Technol 80:53–63.
Ma, F., Hanna, M.A. 1999. Biodiesel production: a review. Bioresour Technol 70:1–15.
Meher, L.C., Vidya Sagar, D., Naik, S.N. 2006. Technical aspects of biodiesel production by transesterification—a review. Renew Sustain Energy Rev 10:248–268.
Mittelbach, M., Gangl, S. 2001. Long storage stability of biodiesel made from rapeseed and used frying oil. JAOCS 78:573–577.
Mittelbach, M., Remschmidt, C. 2004. Biodiesels–The Comprehensive Handbook. Karl-Franzens University, Graz, Austria.
Mudge, S.M., Pereira, G. 1999. Stimulating the biodegradation of crude oil with biodiesel: preliminary results. Spill Sci Technol Bull 5:353–355.
Nitschke, W.R., Wilson, C.M. 1965. Rudolph Diesel, Pioneer of the Age of Power. University of Oklahoma Press, Norman, OK.
Pinto, A.C., Guarieiro, L.L.N., Rezende, M.J.C., Ribeiro, N.M., Torres, E.A., Lopes, W.A., Pereira, P.A.P., Andrade, J.B. 2005. Biodiesel: an overview. J Brazil Chem Soc 16:1313–1330.
Puhan, S., Vedaraman, N., Rambrahaman, B.V., Nagarajan, G. 2005. Mahua (Madhuca indica) seed oil: a source of renewable energy in India. J Sci Ind Res 64:890–896.
Reijnders, L. 2006. Conditions for the sustainability of biomass based fuel use. Energy Policy 34:863–876.
Sastry, G.S.R., Krishna Murthy, A.S.R., Ravi Prasad, P., Bhuvaneswari, K., Ravi, P.V. 2006. Identification and determination of bio-diesel in Diesel. Energy Sources Part A 28: 1337–1342.
Sensoz, S., Angin, D., Yorgun, S. 2000. Influence of particle size on the pyrolysis of rapeseed (Brassica napus L.): fuel properties of bio-oil. Biomass Bioenergy 19:271–279.
Sheehan, J., Dunahay, T., Benemann, J., Roessler, P. 1998. A Look Back at the U.S. Department of Energy's Aquatic Species Program—Biodiesel from Algae. National Renewable Energy Laboratory (NREL) Report: NREL/TP-580-24190. Golden, CO.
Speidel, H.K., Lightner, R.L., Ahmed, I. 2000. Biodegradability of new engineered fuels compared to conventional petroleum fuels and alternative fuels in current use. Appl Biochem Biotechnol 84-86:879–897.
Venkataraman, N.S (ed.) 2002. Focus on bio-diesel. Nandini Chem J IX(10):19–21.
Wardle, D.A. 2003. Global sale of green air travel supported using biodiesel. Renew Sust Energy Rev 7:1–64.
Zhang, Y., Dub, M.A., McLean, D.D., Kates, M. 2003. Biodiesel production from waste cooking oil: 2. Economic assessment and sensitivity analysis. Bioresour Technol 90:229–240.

Chapter 5
Biodiesel from Triglycerides via Transesterification

5.1 Biodiesel from Triglycerides via Transesterification

The possibility of using vegetable oils as fuel has been recognized since the beginning of diesel engines. Vegetable oil has too high a viscosity for use in most existing diesel engines as a straight replacement fuel oil. There are a number of ways to reduce vegetable oils' viscosity. Dilution, microemulsification, pyrolysis, and transesterification are the four techniques applied to solve the problems encountered with high fuel viscosity. One of the most common methods used to reduce oil viscosity in the biodiesel industry is called transesterification. Chemical conversion of the oil into its corresponding fatty ester is called transesterification (Bala, 2005).

Transesterification (also called alcoholysis) is the reaction of a fat or oil triglyceride with an alcohol to form esters and glycerol. Figure 5.1 shows the transesterification reaction of triglycerides. A catalyst is usually used to improve the reaction rate and yield. Because the reaction is reversible, excess alcohol is used to shift the equilibrium to the product side.

Figure 5.2 shows enzymatic biodiesel production by interesterification with methyl acetate in the presence of lipase enzyme as catalyst.

The biodiesel reaction requires a catalyst such as sodium hydroxide to split the oil molecules and an alcohol (methanol or ethanol) to combine with the separated

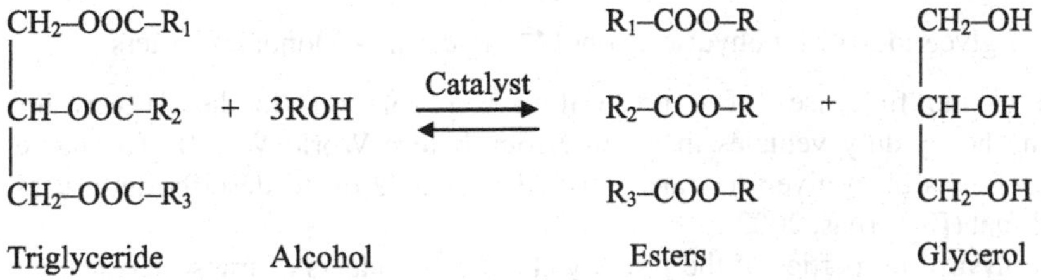

$$\begin{array}{lcccccc} CH_2\text{–}OOC\text{–}R_1 & & & & R_1\text{–}COO\text{–}R & & CH_2\text{–}OH \\ | & & & & & & | \\ CH\text{–}OOC\text{–}R_2 & + & 3ROH & \underset{}{\overset{\text{Catalyst}}{\rightleftarrows}} & R_2\text{–}COO\text{–}R & + & CH\text{–}OH \\ | & & & & & & | \\ CH_2\text{–}OOC\text{–}R_3 & & & & R_3\text{–}COO\text{–}R & & CH_2\text{–}OH \\ \text{Triglyceride} & & \text{Alcohol} & & \text{Esters} & & \text{Glycerol} \end{array}$$

Fig. 5.1 Transesterification of triglycerides with alcohol

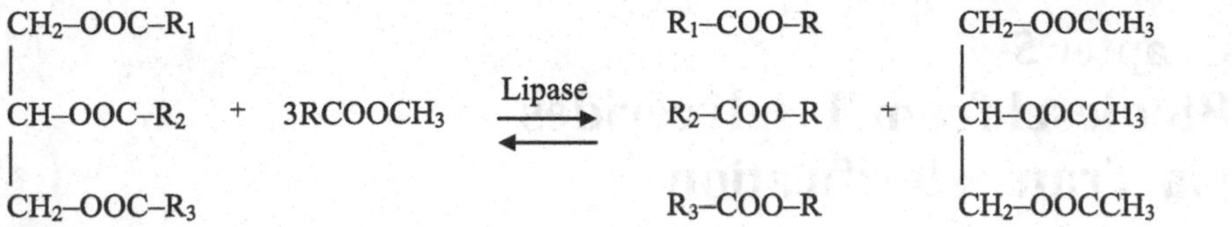

$$\begin{array}{l} CH_2\text{–}OOC\text{–}R_1 \\ | \\ CH\text{–}OOC\text{–}R_2 \\ | \\ CH_2\text{–}OOC\text{–}R_3 \end{array} + 3RCOOCH_3 \underset{}{\overset{Lipase}{\rightleftarrows}} \begin{array}{l} R_1\text{–}COO\text{–}R \\ \\ R_2\text{–}COO\text{–}R \\ \\ R_3\text{–}COO\text{–}R \end{array} + \begin{array}{l} CH_2\text{–}OOCCH_3 \\ | \\ CH\text{–}OOCCH_3 \\ | \\ CH_2\text{–}OOCCH_3 \end{array}$$

Fig. 5.2 Enzymatic biodiesel production by interesterification with methyl acetate

esters. The main byproduct is glycerine. The process reduces the viscosity of the end product. Transesterification is widely used to reduce vegetable oil viscosity (Pinto *et al.*, 2005). Biodiesel is a renewable fuel source. It can be produced from oil from plants or from animal fats that are byproducts in meat processing.

One popular process for producing biodiesel from fats/oils is *trans*esterification of triglyceride by methanol (methanolysis) to make methyl esters of straight-chain fatty acids. The purpose of the transesterification process is to lower the viscosity of oil. The transesterification reaction proceeds well in the presence of some homogeneous catalysts such as potassium hydroxide (KOH) and sodium hydroxide (NaOH) and sulfuric acid or heterogeneous catalysts such as metal oxides or carbonates. Sodium hydroxide is very well accepted and widely used because of its low cost and high product yield (Demirbas, 2003).

Transesterification is the general term used to describe the important class of organic reactions where one ester is transformed into another through interchange of the alkoxy moiety. When the original ester is reacted with an alcohol, the transesterification process is called alcoholysis. In this text, the term transesterification will be used as a synonym of alcoholysis of carboxylic esters, in agreement with most publications in this field. Transesterification is an equilibrium reaction and the transformation occurs essentially by mixing the reactants. However, the presence of a catalyst accelerates considerably the adjustment of the equilibrium. To achieve a high yield of the ester, the alcohol has to be used in excess. Transesterification is the process of exchanging the alkoxy group of an ester compound by another alcohol. These reactions are often catalyzed by the addition of a base and acid. Bases can catalyze the reaction by removing a proton from the alcohol, thus making it more reactive, while acids can catalyze the reaction by donating a proton to the carbonyl group, thus making it more reactive (Schuchardt *et al.*, 1998). The transesterification reaction proceeds with a catalyst or without any catalyst by using primary or secondary monohydric aliphatic alcohols having one to eight carbon atoms as follows:

$$\text{Triglycerides} + \text{Monohydric alcohol} \leftrightarrows \text{Glycerine} + \text{Monoalkyl esters}$$

One of the first uses of transesterified vegetable oil (biodiesel) was for powering heavy-duty vehicles in South Africa before World War II. The name "biodiesel" has been given to transesterified vegetable oil to describe its use as a diesel fuel (Demirbas, 2002).

The physical properties of the primary chemical products of transesterification are given in Table 5.1. The high viscosity of vegetable oils was the cause of severe operational problems such as engine deposits. This is a major reason why neat vegetable oils have largely been abandoned as alternative diesel fuels in favor of monoalkyl esters such as methyl esters. A comparison of various methanolic transesterification methods is presented in Table 5.2.

Table 5.1 Physical properties of chemicals related to transesterification

Name	Specific gravity (g/mL)	Melting point (K)	Boiling Point (K)	Solubility (<10%)
Methyl myristate	0.875	291.0	–	–
Methyl palmitate	0.825	303.8	469.2	Benzene, EtOH, Et_2O
Methyl stearate	0.850	311.2	488.2	Et_2O, chloroform
Methyl oleate	0.875	253.4	463.2	EtOH, Et_2O
Methanol	0.792	176.2	337.9	H_2O, ether, EtOH
Ethanol	0.789	161.2	351.6	H_2O, ether
Glycerol	1.260	255.3	563.2	H_2O, ether

Table 5.2 Comparison of various methanolic transesterification methods

Method	Reaction temperature (K)	Reaction time (min)
Acid or alkali catalytic process	303–338	60–360
Boron trifluoride–methanol	360–390	20–50
Sodium methoxide–catalyzed	293–298	4–6
Non-catalytic supercritical methanol	523–573	6–12
Catalytic supercritical methanol	523–573	0.5–1.5

5.1.1 Catalytic Transesterification Methods

Vegetable oils can be transesterified by heating them with a large excess of anhydrous methanol and a catalyst. The transesterification reaction can be catalyzed by alkalis (Gryglewicz, 1999; Zhang *et al.*, 2003), acids (Furuta *et al.*, 2004), or enzymes (Shieh *et al.*, 2003; Hama *et al.*, 2004; Oda *et al.*, 2004; Du *et al.*, 2004; Noureddini *et al.*, 2005). Various studies have been carried out using different oils as raw material, different alcohols (methanol, ethanol, butanol), as well as different catalysts, including homogeneous ones such as sodium hydroxide, potassium hydroxide, sulfuric acid, and supercritical fluids and heterogeneous ones such as lipases (Marchetti *et al.*, 2007).

5.1.1.1 Acid-catalyzed Transesterification Methods

Sulfuric acid, hydrochloric acid, and sulfonic acid are usually preferred as acid catalysts. The catalyst is dissolved into methanol by vigorous stirring in a small reactor. The oil is transferred into the biodiesel reactor and then the catalyst/alcohol mixture is pumped into the oil.

Methanolic Hydrogen Chloride

Transesterification is carried out with an acidic reagent that is 5% (w/v) anhydrous hydrogen chloride in methanol. It is most often prepared by bubbling hydrogen chloride gas into dry methanol. The hydrogen chloride gas is commercially available in cylinders or can be prepared by dropping concentrated sulfuric acid slowly onto fused ammonium chloride or into concentrated hydrochloric acid. This method is best suited to bulk preparation of the reagent. The hydrogen chloride gas can be obtained by adding acetyl chloride (5 mL) slowly to cooled dry methanol (50 mL).

Methanolic Sulfuric Acid

Vegetable oils are transesterified very rapidly by heating in 10% sulfuric acid in methanol until reflux. A solution of 1 to 2% concentrated sulfuric acid in methanol has almost identical properties to 5% methanolic hydrogen chloride and is very easy to prepare.

Boron Trifluoride Methanol

Boron-trifluoride-catalyzed transesterification of vegetable oils is one of the most popular methods. For transesterification of vegetable oils boron trifluoride (BF_3) is used in methanol (15 to 20% w/v).

5.1.1.2 Alkali Catalytic Transesterification Methods

In the alkali catalytic methanol transesterification method, the catalyst (KOH or NaOH) is dissolved into methanol by vigorous stirring in a small reactor. The oil is transferred into a biodiesel reactor and then the catalyst/alcohol mixture is pumped into the oil. The final mixture is stirred vigorously for 2 h at 340 K in ambient pressure. A successful transesterification reaction produces two liquid phases: ester and crude glycerine.

Crude glycerine, the heavier liquid, will collect at the bottom after several hours of settling. Phase separation can be observed within 10 min and can be complete within 2 h of settling. Complete settling can take as long as 20 h. After settling is complete, water is added at the rate of 5.5% by volume of the methyl ester of oil and then stirred for 5 min, and the glycerine is allowed to settle again. Washing the ester is a two-step process carried out with extreme care. A water wash solution at the rate of 28% by volume of oil and 1 g of tannic acid per liter of water is added to the ester and gently agitated. Air is carefully introduced into the aqueous layer while simultaneously stirring very gently. This process is continued until the ester layer becomes clear. After settling, the aqueous solution is drained and water alone is added at 28% by volume of oil for the final washing (Ma and Hanna, 1999; Demirbas, 2002; Acaroglu and Demirbas, 2007). The resulting biodiesel fuel when used directly in a diesel engine will burn up to 75% cleaner than petroleum D2 fuel.

Sodium-methoxide-catalyzed Transesterification

For sodium-methoxide-catalyzed transesterification, 100 g of vegetable oil is transesterified in toluene (80 mL) and methanol (200 mL) containing fresh sodium (0.8 g) in 10 min at reflux, and a related procedure is used to transesterify liter quantities of oils.

5.1.1.3 Methylation of Free Fatty Acids with Diazomethane (CH_2N_2)

CH_2N_2 reacts rapidly with free fatty acids to give methyl esters. The CH_2N_2 is generally prepared in ethereal solution by the action of an alkali (a 30% solution of KOH) on a nitrosamide, *e.g.*, *N*-methyl-*N*-nitroso-*p*-toluene-sulfonamide or nitroso-methyl-urea (Schelenk and Gellerman, 1960; Demirbas, 1991).

5.1.2 Supercritical Alcohol Transesterification

In general, methyl and ethyl alcohols are used in supercritical alcohol transesterification. In the conventional transesterification of animal fats and vegetable oils for biodiesel production, free fatty acids and water always produce negative effects since the presence of free fatty acids and water causes soap formation, consumes the catalyst, and reduces catalyst effectiveness, all of which results in a low conversion (Komers *et al.*, 2001). The transesterification reaction may be carried out using either basic or acidic catalysts, but these processes require relatively time-consuming and complicated separation of the product and the catalyst, which results in high production costs and energy consumption. To overcome these problems, Saka and Kusdiana (2001) and Demirbas (2002, 2003) have proposed that biodiesel fuels may be prepared from vegetable oil via non-catalytic transesterification with supercritical methanol (SCM). A novel process of biodiesel fuel production has been developed by a non-catalytic supercritical methanol method. Supercritical methanol is believed to solve the problems associated with the two-phase nature of normal methanol/oil mixtures by forming a single phase as a result of the lower value of the dielectric constant of methanol in the supercritical state. As a result, the reaction was found to be complete in a very short time. Compared with the catalytic processes under barometric pressure, the supercritical methanol process is non-catalytic, involves a much simpler purification of products, has a lower reaction time, is more environmentally friendly, and requires lower energy use. However, the reaction requires temperatures of 525 to 675 K and pressures of 35 to 60 MPa (Demirbas, 2003; Kusdiana and Saka, 2001).

5.1.2.1 Non-catalytic Supercritical Methanol Transesterification

Non-catalytic supercritical methanol transesterification is performed in a stainless steel cylindrical reactor (autoclave) at 520 K (Demirbas, 2002).

In a typical run, the autoclave is charged with a given amount of vegetable oil and liquid methanol with changed molar ratios. After each run, the gas is vented and the autoclave is poured into a collecting vessel. The rest of the contents are removed from the autoclave by washing with methanol.

The most important variables affecting the methyl ester yield during transesterification reaction are molar ratio of alcohol to vegetable oil and reaction temperature. Viscosities of the methyl esters from the vegetable oils were slightly higher than that of D2 fuel.

In the transesterification process, the vegetable oil should have an acid value of less than 1 and all materials should be substantially anhydrous. If the acid value is greater than 1, more NaOH or KOH is spent to neutralize the free fatty acids. Water also causes soap formation and frothing (Demirbas, 2003).

The stoichiometric ratio for transesterification reaction requires three moles of alcohol and one mole of triglyceride to yield three moles of fatty acid ester and one mole of glycerol. Higher molar ratios result in greater ester production in a shorter time. In one study, the vegetable oils were transesterified at 1:6 to 1:40 vegetable oil-alcohol molar ratios in catalytic and supercritical alcohol conditions (Demirbas, 2002).

Table 5.3 shows critical temperatures and critical pressures of various alcohols. Table 5.4 shows the comparisons between the catalytic methanol method and the supercritical methanol method for biodiesel from vegetable oils by transesterification. The supercritical methanol process is non-catalytic, involves simpler purification, has a lower reaction time, and is less energy intensive. Therefore, the

Table 5.3 Critical temperatures and critical pressures of various alcohols

Alcohol	Critical temperature (K)	Critical pressure (MPa)
Methanol	512.2	8.1
Ethanol	516.2	6.4
1-Propanol	537.2	5.1
1-Butanol	560.2	4.9

Table 5.4 Comparisons between catalytic methanol (MeOH) method and supercritical methanol (SCM) method for biodiesel from vegetable oils by transesterification

	Catalytic MeOH process	SCM method
Methylating agent	Methanol	Methanol
Catalyst	Alkali (NaOH or KOH)	None
Reaction temperature (K)	303–338	523–573
Reaction pressure (MPa)	0.1	10–25
Reaction time (min)	60–360	7–15
Methyl ester yield (wt.%)	96	98
Removal for purification	Methanol, catalyst, glycerol, soaps	Methanol
Free fatty acids	Saponified products	Methyl esters, water
Smelling from exhaust	Soap smell	Sweet smelling

supercritical methanol method would be more effective and efficient than the common commercial process (Kusdiana and Saka, 2004).

The parameters affecting methyl ester formation are reaction temperature, pressure, molar ratio, water content, and free fatty acid content. It is evident that at subcritical states of alcohol, the reaction rate is so low and gradually increased as either pressure or temperature rises. It was observed that increasing the reaction temperature, especially to supercritical conditions, had a favorable influence on the yield of ester conversion. The yield of alkyl ester increased when the molar ratio of oil to alcohol was increased (Demirbas, 2002). In the supercritical alcohol transesterification method, the yield of conversion rises 50 to 95% for the first 10 min. A simple autoclave with a lab scale for supercritical alcohol transesterification is shown in Fig. 5.3. Figure 5.4 shows the plots for changes in fatty acids alkyl esters conversion from triglycerides as treated in supercritical alcohols at 575 K.

Water content is an important factor in the conventional catalytic transesterification of vegetable oil. In the conventional transesterification of fats and

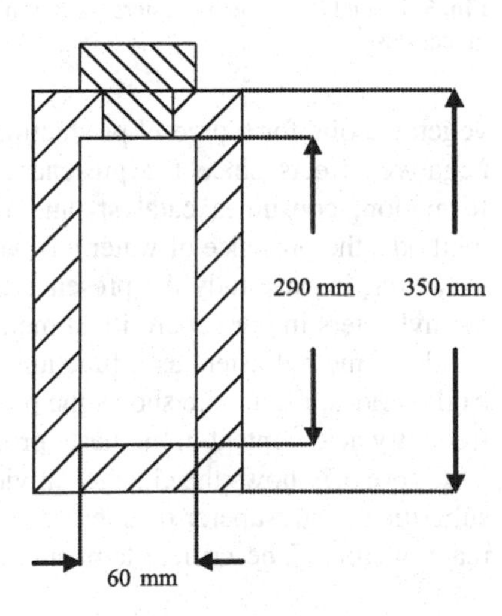

Fig. 5.3 Simple autoclave for supercritical transesterification

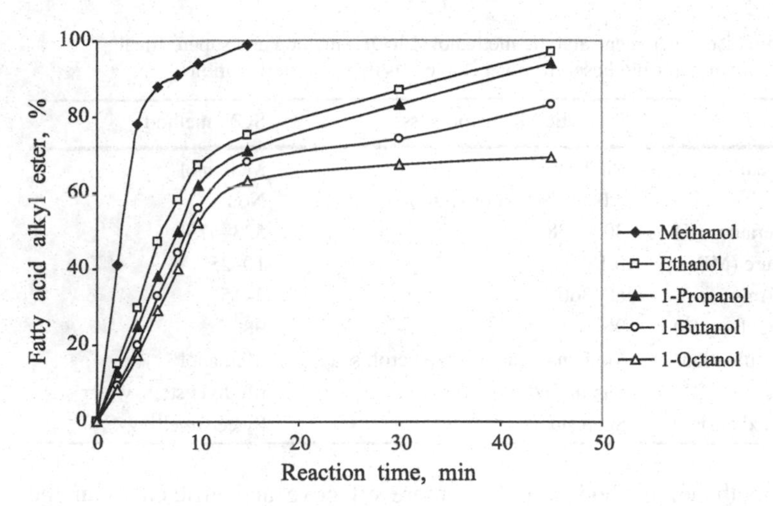

Fig. 5.4 Changes in fatty acid alkyl ester conversion from triglycerides as treated in supercritical alcohol at 575 K

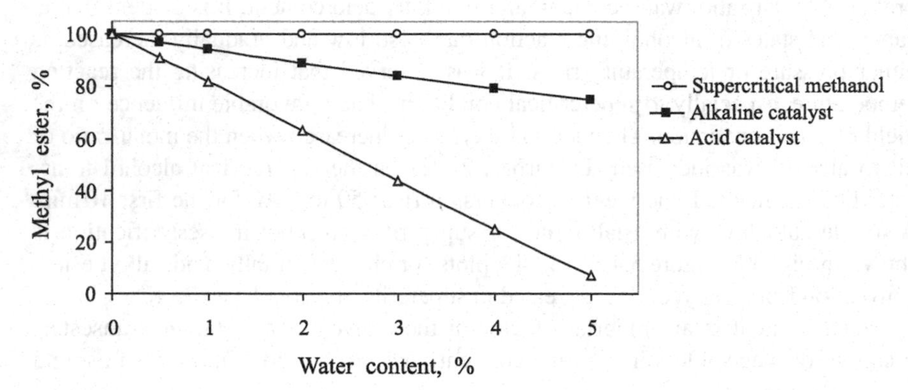

Fig. 5.5 Yields of methyl esters as a function of water content in transesterification of triglycerides

vegetable oils for biodiesel production, free fatty acids and water always produce negative effects since the presence of free fatty acids and water causes soap formation, consumes catalyst, and reduces catalyst effectiveness. In catalyzed methods, the presence of water has negative effects on the yields of methyl esters. However, in one study the presence of water affected positively the formation of methyl esters in our supercritical methanol method. Figure 5.5 shows the plots for yields of methyl esters as a function of water content in the transesterification of triglycerides. Figure 5.6 shows the plots for yields of methyl esters as a function of free fatty acid content in biodiesel production (Kusdiana and Saka, 2004).

Figure 5.7 shows the changes in yield percentage of ethyl esters as treated with subcritical and supercritical ethanol at different temperatures as a function of reaction time. The critical temperature and the critical pressure of methanol are

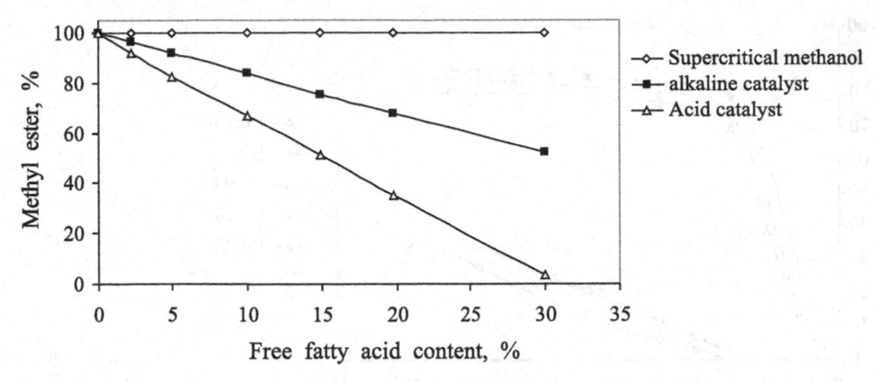

Fig. 5.6 Yields of methyl esters as a function of free fatty acid content in biodiesel production

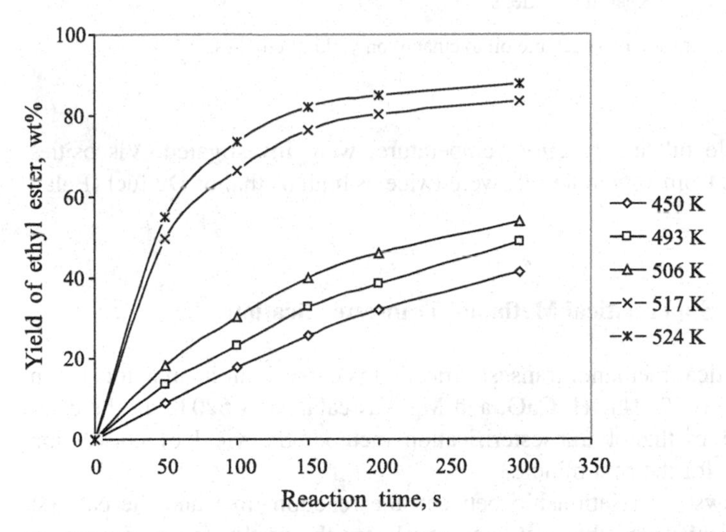

Fig. 5.7 Changes in yield percentage of ethyl esters as treated with subcritical and supercritical ethanol at different temperatures as a function of reaction time. Molar ratio of vegetable oil to ethyl alcohol: 1:40

516.2 K and 6.4 MPa, respectively. It was observed that increasing reaction temperatures, especially supercritical temperatures, had a favorable influence on ester conversion.

Figure 5.8 shows the effect of the molar ratio of sunflower seed oil to ethanol on the yield of ethyl esters at 517 K. The sunflower seed oil was transesterified 1:1, 1:3, 1:9, 1:20, and 1:40 vegetable oil-ethanol molar ratios in supercritical ethanol conditions. It was observed that increasing molar ratio had a favorable influence on ester conversion (Balat, 2005).

Ethyl esters of vegetable oils have several outstanding advantages among other new-renewable and clean-engine fuel alternatives. The variables affecting the ethyl ester yield during transesterification reaction, such as the molar ratio of al-

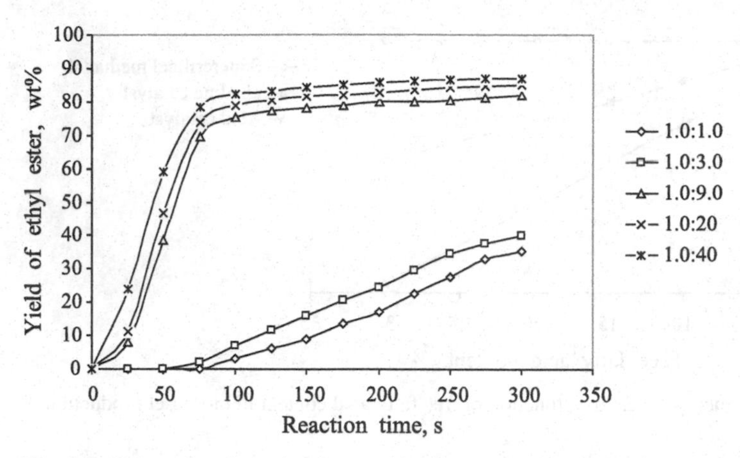

Fig. 5.8 Effect of molar ratio of vegetable oil to ethanol on yield of ethyl ester
Temperature: 517 K

cohol to vegetable oil and reaction temperature, were investigated. Viscosities of the ethyl esters from vegetable oils were twice as high as that of D2 fuel (Balat, 2005).

5.1.2.2 Catalytic Supercritical Methanol Transesterification

Catalytic supercritical methanol transesterification is carried out in an autoclave in the presence of 1 to 5% NaOH, CaO, and MgO as catalyst at 520 K. In the catalytic supercritical methanol transesterification method, the yield of conversion rises to 60 to 90% for the first minute.

Figure 5.9 shows the relationship between the reaction time and the catalyst content. It can be affirmed that CaO can accelerate the methyl ester conversion from sunflower oil at 525 K and 24 MPa even if a small amount of catalyst (0.3% of the oil) was added. The transesterification speed obviously improved as the content of CaO increased from 0.3% to 3%. However, further enhancement of CaO content to 5% produced little increase in methyl ester yield.

Figures 5.10 and 5.11 show the relationships between the temperature and methyl ester yield of non-catalytic and catalytic (3% CaO) transesterifications in subcritical and supercritical methanol from sunflower oil. It was observed that increasing the reaction temperature had a favorable influence on the yield of methyl esters with or without CaO. As shown in Figs. 5.9 and 5.10, at temperatures of 465 K, the yields of methyl esters are relatively low even after reaction for 1200 s and 1600 s, and the yields of methyl esters with or without CaO are only 64.7% and 29.3%, respectively. As the temperature increased, the yield improved significantly. In the 3% CaO catalytic run, the transesterification reaction was essentially completed within 1200 s and 1000 s at 215 K and 225 K, respectively (Demirbas, 2007).

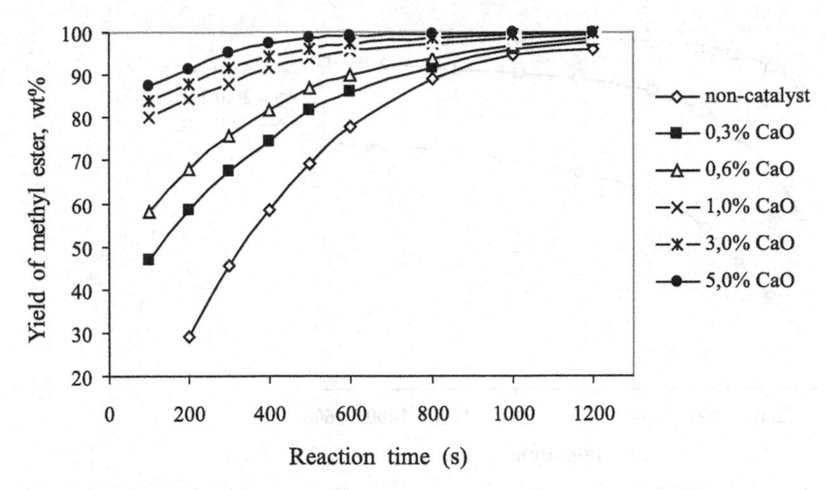

Fig. 5.9 Effect of CaO content on methyl ester yield. Temperature: 525 K; molar ratio of methanol to sunflower oil: 41:1

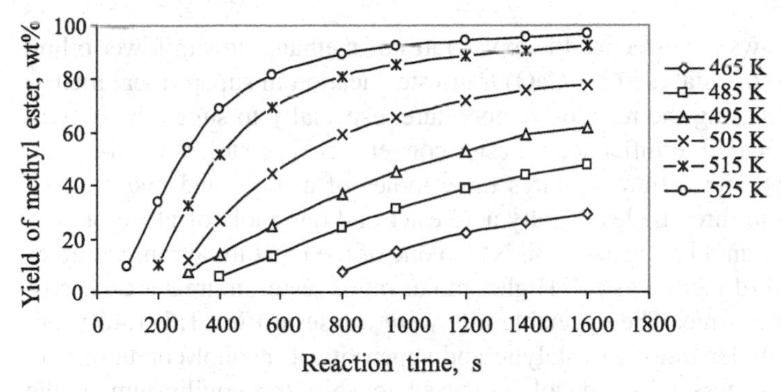

Fig. 5.10 Effect of temperature on methyl ester yield of non-catalytic transesterification in sub- and supercritical methanol from sunflower oil

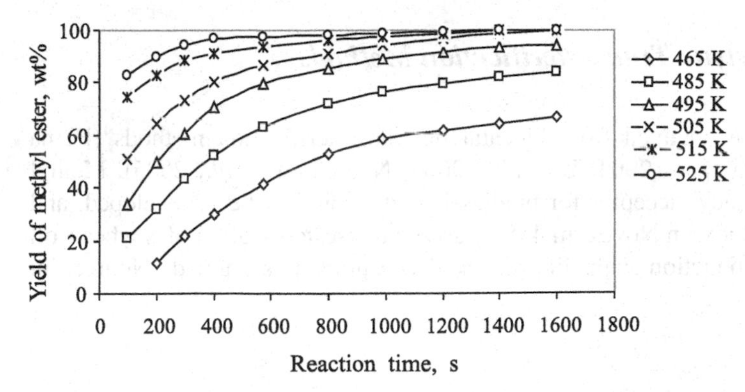

Fig. 5.11 Effect of temperature on methyl ester yield of catalytic (3% CaO) transesterification in sub- and supercritical methanol from sunflower oil

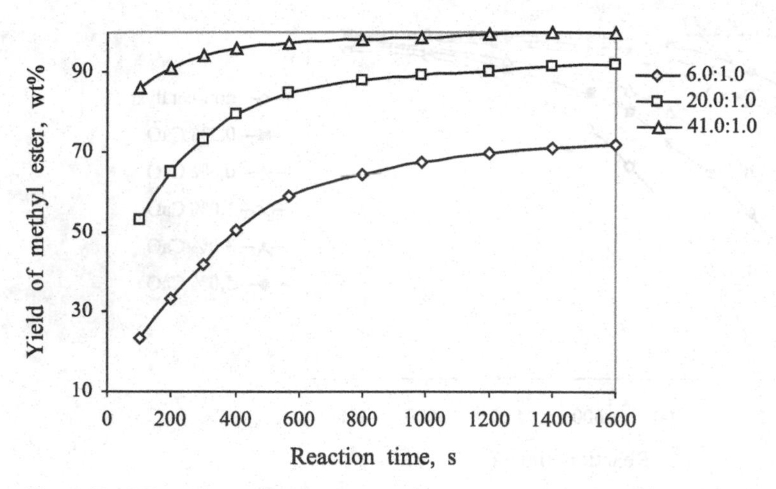

Fig. 5.12 Effect of molar ratio of methanol to sunflower oil on methyl ester yield of catalytic (3% CaO) transesterification in supercritical methanol at 525 K

Figure 5.12 shows the effect of the molar ratio of methanol to sunflower oil on methyl ester yield of catalytic (3% CaO) transesterification in supercritical methanol at 525 K. Increasing the reaction temperature, especially to supercritical temperatures, had a favorable influence on ester conversion. The stoichiometric ratio for transesterification reactions requires three moles of alcohol and one mole of triglyceride to yield three moles of fatty acid ester and one mole of glycerol. The molar ratio of methanol to vegetable oil is also one of the most important variables affecting the yield of methyl esters. Higher molar ratios result in greater ester production in a shorter time. The vegetable oils were transesterified 1:6–1:40 vegetable oil-alcohol molar ratios in catalytic and supercritical alcohol conditions. In this reaction, an excess of methanol was used to shift the equilibrium in the direction of the products (Demirbas, 2007).

5.1.3 Biocatalytic Transesterification Methods

Biodiesel can be obtained from biocatalytic transesterification methods (Hama *et al.*, 2004; Oda *et al.*, 2004; Du *et al.*, 2004; Noureddini *et al.*, 2005). Methyl acetate, a novel acyl acceptor for biodiesel production, has been developed, and a comparative study on Novozym 435-catalyzed transesterification of soybean oil for biodiesel production with different acyl acceptors was studied (Noureddini *et al.*, 2005).

5.1.4 Recovery of Glycerine

The ASTM standards ensure that the following important factors in the biodiesel production process by transesterification are satisfied: (a) complete transesterification reaction, (b) removal of catalyst, (c) removal of alcohol, (d) removal of glycerol, and (e) complete esterification of free fatty acids. The following transesterification procedure is for methyl ester production. The catalyst is dissolved into the alcohol by vigorous stirring in a small reactor. The oil is transferred into the biodiesel reactor, and then the catalyst/methanol mixture is pumped into the oil and the final mixture stirred vigorously for 2 h. A successful reaction produces two liquid phases: ester and crude glycerol. The entire mixture then settles and glycerol is left on the bottom and a methyl ester (biodiesel) is left on top. Crude glycerol, the heavier liquid, will collect at the bottom after several hours of settling. Phase separation can be observed within 10 min and can be complete within 2 h after stirring has stopped. Complete settling can take as long as 18 h. After settling is complete, water is added at the rate of 5.0% by volume of the oil and then stirred for 5 min and the glycerol allowed to settle again. After settling is complete the glycerol is drained and the ester layer remains (Bala, 2005).

The recovery of high-quality glycerol as a biodiesel byproduct is a primary action to be considered to lower the cost of biodiesel. By neutralizing the free fatty acids, removing the glycerol, and creating an alcohol ester, transesterification occurs. This is accomplished by mixing methanol with sodium hydroxide to make sodium methoxide. This dangerous liquid is then mixed with vegetable oil. Washing the methyl ester is a two-step process that is carried out with extreme care.. This procedure is continued until the methyl ester layer becomes clear. After settling, the aqueous solution is drained and water alone is added at 28% by volume of oil for the final washing. The resulting biodiesel fuel, when used directly in a diesel engine, will burn up to 75% cleaner than D2 fuel (Bala, 2005).

The process of converting vegetable oil into biodiesel fuel is called transesterification and is fortunately less complex than it sounds. Chemically, transesterification means taking a triglyceride molecule or a complex fatty acid, neutralizing the free fatty acids, removing the glycerine, and creating an alcohol ester. This is accomplished by mixing methanol with sodium hydroxide to make sodium methoxide. This dangerous liquid is then mixed into vegetable oil. The entire mixture then settles. Glycerine is left on the bottom and methyl ester, or biodiesel, is left on top. The glycerine can be used to make soap (or any one of 1,600 other products) and the methyl ester is washed and filtered.

5.1.5 General Reaction Mechanism of Transesterification

Triacylglycerols (vegetable oils and fats) are esters of long-chain carboxylic acids combined with glycerol. Carboxylic acids {R– C(=O) – O – H} can be converted into methyl esters {R– C(=O) – O – CH3} by the action of a transesterification agent. The parameters affecting the methyl ester formation are reaction temperature, pressure, molar ratio, water content, and free fatty acid content. It was observed that increasing the reaction temperature had a favorable influence on the yield of ester conversion. The yield of alkyl ester increased when the oil-to-alcohol molar ratio was increased (Demirbas, 2002).

$$\text{Fatty acid } (R_1COOH) + \text{Alcohol } (ROH) \leftrightarrows \text{Ester } (R_1COOR) + \text{Water } (H_2O) \quad (5.1)$$

$$\text{Triglyceride} + ROH \leftrightarrows \text{Diglyceride} + RCOOR_1 \quad (5.2)$$

$$\text{Diglyceride} + ROH \leftrightarrows \text{Monoglyceride} + RCOOR_2 \quad (5.3)$$

$$\text{Monoglyceride} + ROH \leftrightarrows \text{Glycerol} + RCOOR_3 \quad (5.4)$$

Transesterification consists of a number of consecutive, reversible reactions (Schwab *et al.*, 1987; Freedman *et al.*, 1986). The triglyceride is converted stepwise into diglyceride, monoglyceride, and, finally, glycerol (Eqs 5.1–5.4) in which 1 mol of alkyl esters is removed in each step. The reaction mechanism for alkali-catalyzed transesterification was formulated as three steps (Eckey, 1956; Sridharan and Mathai, 1974). The formation of alkyl esters from monoglycerides is believed to be the step that determines the reaction rate since monoglycerides are the most stable intermediate compound (Ma and Hanna, 1999).

Several aspects, including the type of catalyst (alkaline, acid, or enzyme), alcohol/vegetable oil molar ratio, temperature, purity of the reactants (mainly water content), and free fatty acid content have an influence on the course of transesterification. In the conventional transesterification of fats and vegetable oils for biodiesel production, free fatty acids and water always produce negative effects (Kusdiana and Saka, 2004)

Transesterification is the general term used to describe the important class of organic reactions where one ester is transformed into another ester through interchange of the alkoxy moiety. When the original ester is reacted with an alcohol, the transesterification process is called alcoholysis (Schuchard *et al.*, 1998). Transesterification is an equilibrium reaction, and the transformation occurs essentially by mixing the reactants. In the transesterification of vegetable oils, a triglyceride reacts with an alcohol in the presence of a strong acid or base, producing a mixture of fatty acids, alkyl esters, and glycerol. The stoichiometric reaction requires 1 mol of a triglyceride and 3 mol of the alcohol. However, an excess of the alcohol is used to increase the yields of the alkyl esters and to allow their phase separation from the glycerol formed.

5.1.5.1 Acid-catalyzed Transesterification

The transesterification process is catalyzed by Brønsted acids, preferably by sulfonic and sulfuric acids. These catalysts give very high yields in alkyl esters, but the reactions are slow. The alcohol/vegetable oil molar ratio is one of the main factors that influence transesterification. An excess of the alcohol favors the formation of alkyl esters. On the other hand, an excessive amount of alcohol makes the recovery of the glycerol difficult, so that the ideal alcohol/oil ratio has to be established empirically, considering each individual process. Figure 5.13 shows the mechanism of acid-catalyzed esterification of fatty acids. The initial step is protonation of the acid to give an oxonium ion (1), which can undergo an exchange reaction with an alcohol to give the intermediate (2), and this in turn can lose a proton to become an ester (3). Each step in the process is reversible, but in the presence of a large excess of the alcohol, the equilibrium point of the reaction is displaced so that esterification proceeds virtually to completion.

Figure 5.14 shows the mechanism of acid-catalyzed transesterification of vegetable oils (Christie, 1989). The transesterification occurs under similar conditions (Fig. 5.3). In this instance, initial protonation of the ester is followed by addition of the exchanging alcohol to give the intermediate (4), which can be dissociated via the transition state (5) to give the ester (6).

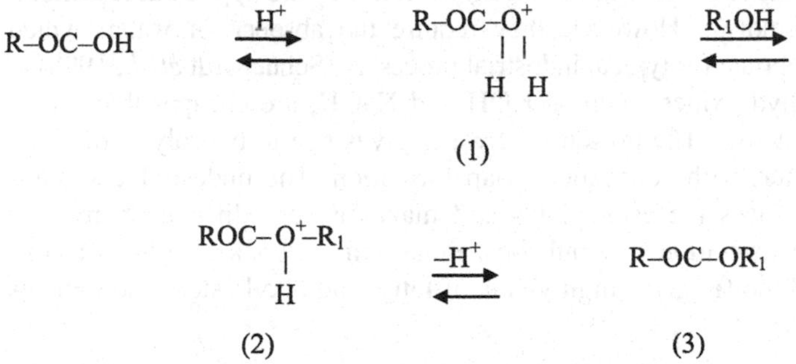

$$\text{R–OC–OH} \underset{}{\overset{H^+}{\rightleftharpoons}} \underset{(1)}{\text{R–OC–O}^+(\text{H})(\text{H})} \overset{R_1OH}{\rightleftharpoons}$$

$$\underset{(2)}{\text{ROC–O}^+(\text{H})\text{–R}_1} \overset{-H^+}{\rightleftharpoons} \underset{(3)}{\text{R–OC–OR}_1}$$

Fig. 5.13 Mechanism of acid-catalyzed esterification of fatty acids

$$\text{R–OC–R}_1 \overset{H^+}{\rightleftharpoons} \underset{(2)}{\text{R–OC–O}^+(\text{H})\text{–R}_1} \overset{R_2OH}{\rightleftharpoons} \underset{(4)}{\left[\text{R–C(O)(OH)–O(H)–R}_1 ;\ \text{O–R}_2,\ \text{H}\right]^+}$$

$$\rightleftharpoons \underset{(6)}{\text{R–OC–O–R}_2} \overset{-H^+}{\rightleftharpoons} \underset{(5)}{\text{R–OC–O}^+(\text{H})\text{–R}_2}$$

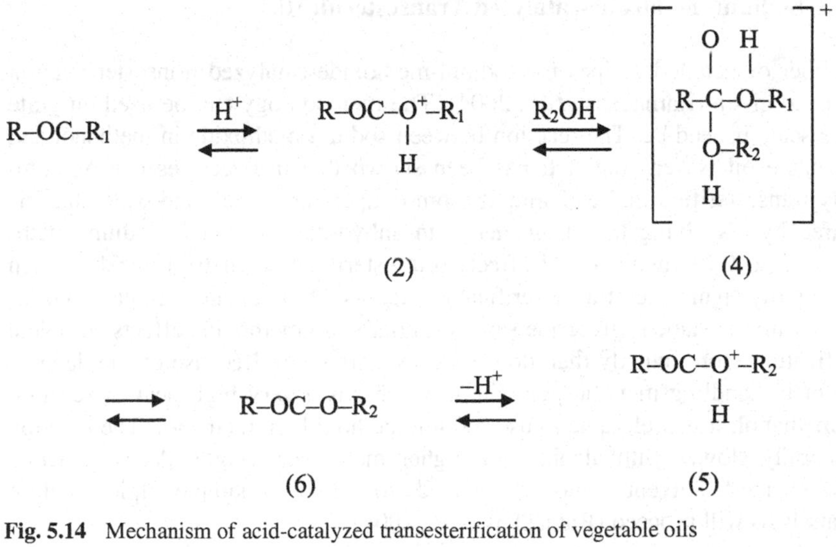

Fig. 5.14 Mechanism of acid-catalyzed transesterification of vegetable oils

5.1.5.2 Base-catalyzed Transesterification

Esters, in the presence of bases such as an alcoholate anion, form an anionic intermediate, which can dissociate back to the original ester or form a new ester. Transesterification can therefore occur by this mechanism with basic catalysis, but esterification cannot.

The base-catalyzed transesterification of vegetable oils proceeds faster than the acid-catalyzed reaction. The first step is the reaction of the base with the alcohol, producing an alkoxide and a protonated catalyst. The nucleophilic attack of the alkoxide at the carbonyl group of the triglyceride generates a tetrahedral intermediate, from which the alkyl ester and the corresponding anion of the diglyceride are formed. The latter deprotonates the catalyst, thus regenerating the active species, which is now able to react with a second molecule of the alcohol, starting another catalytic cycle. Diglycerides and monoglycerides are converted by the same mechanism into a mixture of alkyl esters and glycerol. Alkaline metal alkoxides (as CH_3ONa for methanolysis) are the most active catalysts since they give very high yields (>98%) in short reaction times (30 min) even if they are applied at low molar concentrations (0.5 mol%). However, they require the absence of water, which makes them inappropriate for typical industrial processes (Schuchardt *et al.*, 1998).

Alkaline metal hydroxides, such as KOH and NaOH, are cheaper than metal alkoxides, but less active. The presence of water gives rise to hydrolysis of some of the produced ester, with consequent soap formation. The undesirable saponification reaction reduces the ester yields and makes it very difficult to recover glycerol due to the formation of emulsions. Potassium carbonate, used in a concentration of 2 or 3 mol%, gives high yields of fatty acid alkyl esters and reduces soap formation.

5.1.5.3 Sodium-methoxide-catalyzed Transesterification

A number of detailed recipes for sodium-methoxide-catalyzed transesterification have been given (Ramadhas *et al.*, 2004). The methodology can be used on quite a large scale if need be. The reaction between sodium methoxide in methanol and a vegetable oil is very rapid. It has been shown that triglycerides can be completely transesterified in 2 to 5 min at room temperature. The methoxide anion is prepared by dissolving the clean metals in anhydrous methanol. Sodium methoxide (0.5 to 2 M) in methanol effects transesterification of triglycerides much more rapidly than other transesterification agents. At equivalent molar concentrations with the same triglyceride samples, potassium methoxide effects complete esterification more quickly than does sodium methoxide. Because of the dangers inherent in handling metallic potassium, which has a very high heat of reaction with methanol, it is preferable to use sodium methoxide in methanol. The reaction is generally slower with alcohols of higher molecular weight. As with acidic catalysis, inert solvents must be added to dissolve simple lipids before methanolysis will proceed (Ramadhas *et al.*, 2004).

5.1.5.4 Boron-trifluoride-catalyzed Transesterification

The Lewis acid boron trifluoride, in the form of its coordination complex with methanol, is a powerful acidic catalyst for the esterification of fatty acids. One of the most popular of all transesterification catalysts is boron trifluoride in methanol (12 to 14% w/v), and in particular it is often used as a rapid means of esterifying free fatty acids. Compared with some of the other acidic catalysts under similar conditions, it does not even appear to be any more rapid in its reaction.

5.1.6 Esterification of Fatty Acids with Diazomethane

Diazomethane (CH_2N_2) reacts rapidly with free fatty acids to give methyl esters but does not affect transesterification of other lipids. The reaction is not instantaneous, however, as has sometimes been assumed, unless a little methanol is present as a catalyst (Schelenk and Gellerman, 1960). Carboxylic acids {R– C(=O) – O – H} can be converted into methyl esters {R– C(=O) – O – CH_3} by the action of CH_2N_2:

$$R-C(=O)-O-H+CH_2N_2 \rightarrow R-C(=O)-O-CH_2-H+N_2. \qquad (5.5)$$

Notice that the diazomethane appears to insert itself between the O and the H of the O–H bond (Eq. 5.5). The high reactivity of diazomethane arises from the fact that it possesses an exceedingly reactive leaving group, the nitrogen molecule (N_2). A nucleophilic substitution reaction on the protonated diazomethane molecule transfers a methyl group to the oxygen atom of the carboxylic acid, while liberating a very stable product (N_2 gas). This process is very favorable energetically, owing to the great stability of N_2.

5.1.7 Non-catalytic Supercritical Alcohol Transesterification

Biodiesel, an alternative diesel fuel, is made from renewable biological sources such as vegetable oils and animal fats by non-catalytic supercritical alcohol transesterification methods (Demirbas, 2003). A non-catalytic biodiesel production route with supercritical methanol has been developed that allows a simple process and high yield because of simultaneous transesterification of triglycerides and methyl esterification of fatty acids (Demirbas, 2002).

The parameters affecting methyl ester formation are reaction temperature, pressure, molar ratio, water content, and free fatty acid content. It is evident that at a subcritical state of alcohol, the reaction rate is so low and gradually increases as either pressure or temperature rises. It has been observed that increasing the reaction temperature, especially to supercritical conditions, has a favorable influence on the yield of ester conversion. The yield of alkyl ester increases when the oil-to-alcohol molar ratio is increased (Demirbas, 2002). In the supercritical alcohol transesterification method, the yield of conversion rises 50 to 95% for the first 10 min.

Water content is an important factor in the conventional catalytic transesterification of vegetable oil. In the conventional transesterification of fats and vegetable oils for biodiesel production, free fatty acids and water always produce negative effects since the presence of free fatty acids and water causes soap formation, consumes catalyst, and reduces catalyst effectiveness. In catalyzed methods, the presence of water has negative effects on the yields of methyl esters. However, the presence of water affected positively the formation of methyl esters in our supercritical methanol method.

In the supercritical alcohol transesterification method, the yield of conversion increases 50 to 95% for the first 8 min. In the catalytic supercritical methanol transesterification method, the yield of conversion increases 60 to 90% for the first minute.

5.1.8 Enzyme-catalyzed Processes

Transesterification can be carried out chemically or enzymatically. In recent work three different lipases (*Chromobacterium viscosum*, *Candida rugosa*, and Porcine pancreas) were screened for a transesterification reaction of jatropha oil in a solvent-free system to produce biodiesel; only lipase from *Chromobacterium viscosum* was found to give appreciable yield (Shah *et al.*, 2004). Immobilization of lipase (*Chromobacterium viscosum*) on Celite-545 enhanced the biodiesel yield to 71% from the 62% yield obtained by using free tuned enzyme preparation with a process time of 8 h at 113 K. Immobilized *Chromobacterium viscosum* lipase can be used for ethanolysis of oil. It was seen that immobilization of lipases and optimization of transesterification conditions resulted in adequate yield of biodiesel in the case of the enzyme-based process (Shah *et al.*, 2004).

Although the enzyme-catalyzed transesterification processes are not yet commercially developed, new results have been reported in recent articles and patents. The common aspects of these studies consist in optimizing the reaction conditions (solvent, temperature, pH, type of microorganism that generates the enzyme, *etc.*) in order to establish suitable characteristics for an industrial application. However, the reaction yields as well as the reaction times are still unfavorable compared to the base-catalyzed reaction systems (Schuchardt *et al.*, 1998). Due to their ready availability and the ease with which they can be handled, hydrolytic enzymes have been widely applied in organic synthesis.

Methyl acetate, a novel acyl acceptor for biodiesel production, has been developed, and a comparative study on Novozym 435-catalyzed transesterification of soybean oil for biodiesel production with different acyl acceptors has been studied (Du *et al.*, 2004).

References

Acaroglu, M., Demirbas, A.1999. Relationships between viscosity and density measurements of biodiesel fuels. Energy Sour Part A 29:705–712.

Bala, B.K.2005. Studies on biodiesels from transformation of vegetable oils for diesel engines. Energy Edu Sci Technol 15:1–43.

Balat, M. 2005. Biodiesel from vegetable oils via transesterification in supercritical ethanol. Energy Edu Sci Technol 16:45–52

Christie, W.W. 1989. Gas Chromatography and Lipids: a Practical Guide. The Oily Press, Dundee.

Demirbas, A. 1999. Fatty and resin acids recovered from spruce wood by supercritical acetone extraction. Holzforschung 45:337–339.

Demirbas, A. 2002. Biodiesel from vegetable oils via transesterification in supercritical methanol. Energy Convers Mgmt 43:2349–56.

Demirbas, A. 2003. Biodiesel fuels from vegetable oils via catalytic and non-catalytic supercritical alcohol transesterifications and other methods: a survey. Energy Convers Mgmt 44:2093–2109.

Demirbas, A. 2007. Biodiesel from sunflower oil in supercritical methanol with calcium oxide. Energy Convers Mgmt 48:2271–2282.

Du, W., Xu, Y., Liu, D., Zeng, J. 2004. Comparative study on lipase-catalyzed transformation of soybean oil for biodiesel production with different acyl acceptors. J Mol Catal B Enzymat 30:125–129.

Eckey, E.W. 1956. Esterification and interesterification. *JAOCS* 33:575–579.

Freedman, B., Butterfield, R.O., Pryde, E.H. 1986. Transesterification kinetics of soybean oil. JAOCS 63:1375–1380.

Furuta, S., Matsuhashi, H., Arata, K. 2004. Biodiesel fuel production with solid superacid catalysis in fixed bed reactor under atmospheric pressure. Catal Commun 5:721–723.

Gryglewicz, S. 1999. Rapeseed oil methyl esters preparation using heterogeneous catalysts. Bioresour Technol 70:249–253.

Hama, S., Yamaji, H., Kaieda, M., Oda, M., Kondo, A., Fukuda, H. 2004. Effect of fatty acid membrane composition on whole-cell biocatalysts for biodiesel-fuel production. Biochem Eng J 21:155–160.

Komers, K., Machek, J., Stloukal, R. 2001. Biodiesel from rapeseed oil and KOH 2. Composition of solution of KOH in methanol as reaction partner of oil. Eur J Lipid Sci Technol 103:359–362.

Kusdiana, D., Saka, S. 2001. Kinetics of transesterification in rapeseed oil to biodiesel fuels as treated in supercritical methanol. Fuel 80:693–698.

Kusdiana, D., Saka, S. 2004. Effects of water on biodiesel fuel production by supercritical methanol treatment. Bioresour Technol 91:289–295.

Ma, F., Hana, M.A. 1999. Biodiesel production: a review. Bioresour Technol 70:1–15.

Marchetti, J.M., Miguel, V.U., Errazu, A.F. 2007. Possible methods for biodiesel production. Renew Sustain Energy Rev 11:1300–1311.

Noureddini, H., Gao, X., Philkana, R.S. 2005. Immobilized Pseudomonas cepacia lipase for biodiesel fuel production from soybean oil. Bioresour Technol 96:769–777.

Oda, M., Kaieda, M., Hama, S., Yamaji, H., Kondo, A., Izumoto, E., Fukuda, H. 2004. Facilitatory effect of immobilized lipase-producing *Rhizopus oryzae* cells on acyl migration in biodiesel-fuel production. Biochem Eng J 23:45–51.

Pinto, A.C., Guarieiro, L.L.N., Rezende, M.J.C., Ribeiro, N.M., Torres, E.A., Lopes, W.A., Pereira, P.A..P., de Andrade, J.B. 2005. Biodiesel: an overview. J Braz Chem Soc 16:1313–1330.

Ramadhas, A.S., Jayaraj, S., Muraleedharan, C. 2004. Use of vegetable oils as I.C. engine fuels—a review. Renew Energy 29:727–742.

Saka, S., Kusdiana, D. 2001. Biodiesel fuel from rapeseed oil as prepared in supercritical methanol. Fuel 80:225–231.

Schelenk, H., Gellerman, J.L. 1960. Esterification of fatty acids with diazomethane on a small scale. Anal Chem 32:1412–1414.

Schuchardt, U., Ricardo Sercheli, R., Vargas, R.M. 1998. Transesterification of vegetable oils: a review. J Braz Chem Soc 9:199–210.

Schwab, A.W., Bagby, M.O., Freedman, B. 1987. Preparation and properties of diesel fuels from vegetable oils. Fuel 66:1372–1378.

Shah, S., Sharma, S., Gupta, M.N. 2004. Biodiesel preparation by lipase-catalyzed transesterification of *Jatropha* oil. Energy Fuels 18:154–159.

Shieh C.-J., Liao, H.-F., Lee, C.-C. 2003. Optimization of lipase-catalyzed biodiesel by response surface methodology. Bioresour Technol 88:103–106.

Sridharan, R., Mathai, I.M. 1974. Transesterification reactions. J Sci Ind Res 33:178–187.

Zhang, Y., Dub, M.A., McLean, D.D., Kates, M. 2003. Biodiesel production from waste cooking oil: 2. Economic assessment and sensitivity analysis. Bioresour Technol 90:229–240.

Chapter 6
Fuel Properties of Biodiesels

Biodiesels are characterized by their viscosity, density, cetane number, cloud and pour points, distillation range, flash point, ash content, sulfur content, carbon residue, acid value, copper corrosion, and higher heating value (HHV). The most important variables affecting the ester yield during the transesterification reaction are the molar ratio of alcohol to vegetable oil and reaction temperature. The viscosity values of vegetable oil methyl esters decrease sharply after transesterification. Compared to D2 fuel, all of the vegetable oil methyl esters are slightly viscous. The flash point values of vegetable oil methyl esters are significantly lower than those of vegetable oils. There is high regression between the density and viscosity values of vegetable oil methyl esters. The relationships between viscosity and flash point for vegetable oil methyl esters are considerably regular.

6.1 Viscosity, Density, and Flash Point

The properties of biodiesel are similar to those of diesel fuels. Viscosity is the most important property of biodiesels since it affects the operation of fuel injection equipment, particularly at low temperatures when an increase in viscosity affects the fluidity of the fuel. High viscosity leads to poorer atomization of the fuel spray and less accurate operation of the fuel injectors. The lower the viscosity of the biodiesel, the easier it is to pump and atomize and achieve finer droplets (Islam *et al.*, 2004). The conversion of triglycerides into methyl or ethyl esters through the transesterification process reduces the molecular weight to one third that of the triglyceride and reduces the viscosity by a factor of about eight. Viscosities show the same trends as temperatures, with the lard and tallow biodiesels higher than the soybean and rapeseed biodiesels. Biodiesels have a viscosity close to that of diesel fuels. As the oil temperature increases its viscosity decreases. Table 6.1 shows some fuel properties of six methyl ester biodiesels given by different researchers.

Table 6.1 Some fuel properties of six methyl ester biodiesels

Source	Viscosity cSt at 313.2 K	Density g/mL at 288.7 K	Cetane number	Reference
Sunflower	4.6	0.880	49	Pischinger *et al.*, 1982
Soybean	4.1	0.884	46	Schwab *et al.*, 1987
Palm	5.7	0.880	62	Pischinger *et al.*, 1982
Peanut	4.9	0.876	54	Srivastava and Prasad, 2000
Babassu	3.6	–	63	Srivastava and Prasad, 2000
Tallow	4.1	0.877	58	Ali *et al.*, 1995

Table 6.2 Viscosity, density, and flash point measurements of nine oil methyl esters

Methyl ester	Viscosity mm^2/s (at 313 K)	Density kg/m^3 (at 288 K)	Flash point K
Cottonseed oil	3.75	870	433
Hazelnut kernel oil	3.59	860	422
Linseed oil	3.40	887	447
Mustard oil	4.10	885	441
Palm oil	3.94	880	431
Rapeseed oil	4.60	894	453
Safflower oil	4.03	880	440
Soybean oil	4.08	885	441
Sunflower oil	4.16	880	439

Table 6.2 shows the viscosity, density, and flash point measurements of nine oil methyl esters. The viscosity, density, and flash point values of methyl esters decrease considerably via the transesterification process.

Vegetable oils can be used as fuel for combustion engines, but their viscosity is much higher than that of common diesel fuel and requires modifications to the engines. The major problem associated with the use of pure vegetable oils as fuels for diesel engines is high fuel viscosity in the compression ignition. Therefore, vegetable oils are converted into their methyl esters (biodiesel) by transesterification. The viscosity values of vegetable oils are between 27.2 and 53.6 mm^2/s, whereas those of vegetable oil methyl esters are between 3.6 and 4.6 mm^2/s. The viscosity values of vegetable oil methyl esters decrease sharply following the transesterification process. The viscosity of D2 fuel is 2.7 mm^2/s at 311 K (Demirbas, 2003). Compared to D2 fuel, all of the vegetable oil methyl esters are slightly viscous.

The flash point values of vegetable oil methyl esters are much lower than those of vegetable oils. An increase in density from 860 to 885 kg/m^3 for vegetable oil methyl esters or biodiesels increases the viscosity from 3.59 to 4.63 mm^2/s, and the increases are highly regular. There is high regression between the density and viscosity values of vegetable oil methyl esters. The relationships between viscosity and flash point for vegetable oil methyl esters are irregular.

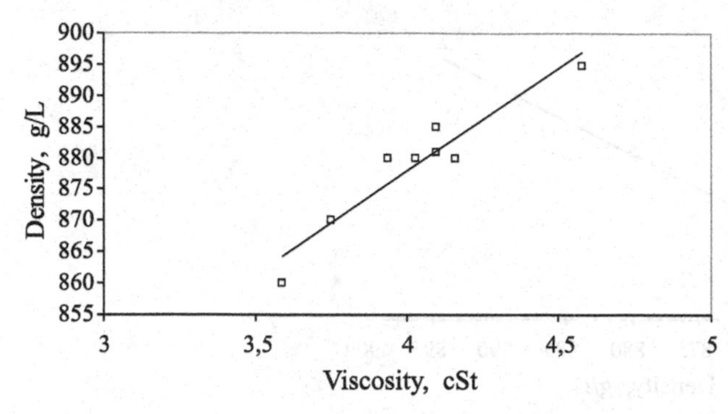

Fig. 6.1 Relationships between viscosity and density for vegetable oil methyl esters

Figure 6.1 shows the relationships between viscosity and density for vegetable oil methyl esters.

$$D = 33.107V + 745.39, \tag{6.1}$$

where *D* is the density and *V* is the viscosity of a biodiesel sample (Eq. 6.1). There is high regression between the viscosity and density values of biodiesel samples ($R^2 = 0.9093$).

Figure 6.2 shows the relationships between the viscosity and the flash point of vegetable oil methyl esters.

$$F = 32.641V + 305.02, \tag{6.2}$$

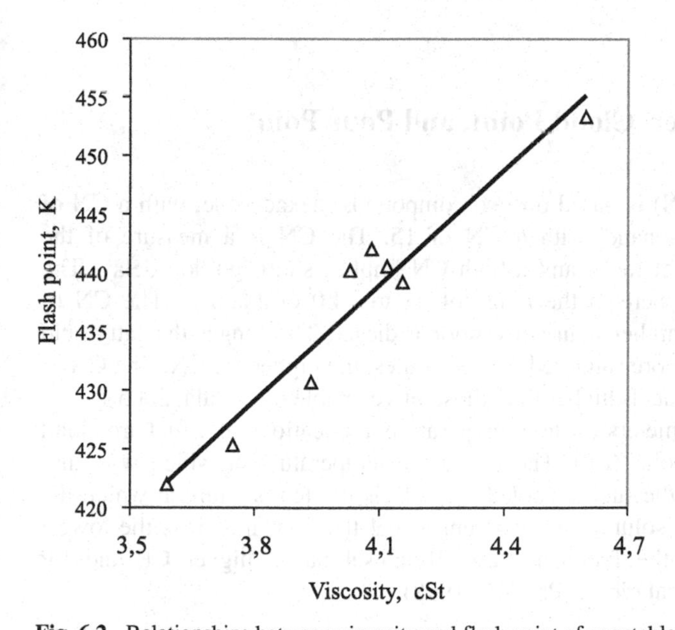

Fig. 6.2 Relationships between viscosity and flash point of vegetable oil methyl esters

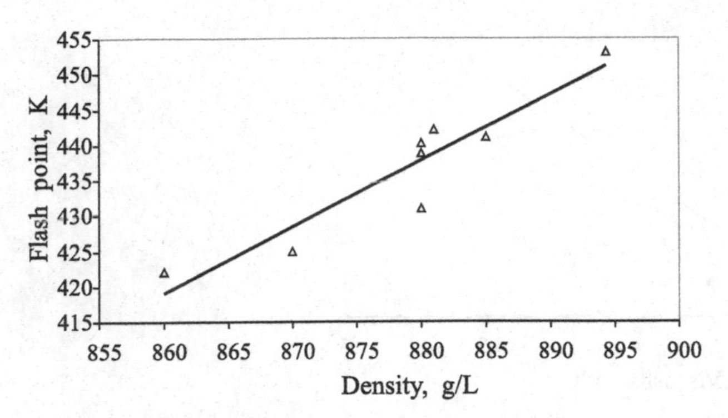

Fig. 6.3 Relationship between density and flash point for vegetable oil methyl esters

where *F* is the flash point and *V* the viscosity of a biodiesel sample (Eq. 6.2). There is high regression between viscosity and density values of biodiesel samples ($R^2 = 0.9357$).

Figure 6.3 shows the relationships between density and flash point for vegetable oil methyl esters.

$$F = 0.9323D - 282.64 \tag{6.3}$$

where *F* is the flash point and *D* the density of a biodiesel sample (Eq. 6.5). There is high regression between the viscosity and density values of biodiesel samples ($R^2 = 0.8753$). There are high regressions between the density, viscosity, and flash point values of vegetable oil methyl esters.

6.2 Cetane Number, Cloud Point, and Pour Point

The cetane number (CN) is based on two compounds, hexadecane, with a CN of 100, and heptamethylnonane, with a CN of 15. The CN is a measure of the ignition quality of diesel fuels, and a high CN implies short ignition delay. The CNs of CSO samples were in the range of 41 to 44.0 (Table 6.4). The CN of biodiesel is generally higher than conventional diesel. The longer the fatty acid carbon chains and the more saturated the molecules, the higher the CN. The CN of biodiesel from animal fats is higher than those of vegetable oils (Bala, 2005).

Two important parameters for low-temperature applications of a fuel are cloud point (CP) and pour point (PP). The CP is the temperature at which wax first becomes visible when the fuel is cooled. The PP is the temperature at which the amount of wax from a solution is sufficient to gel the fuel; thus it is the lowest temperature at which the fuel can flow. Biodiesel has a higher CP and PP compared to conventional diesel (Prakash, 1998).

6.3 Characteristics of Distillation Curves

The distillation curves of biodiesel samples are determined by a test method (ASTM D2887-97) by which the boiling range distribution of liquid fuels is determined.

Figure 6.4 shows the distillation curves of D2 fuel and biodiesel from cottonseed oil. As seen in Fig. 6.4, the most volatile fuel is D2.

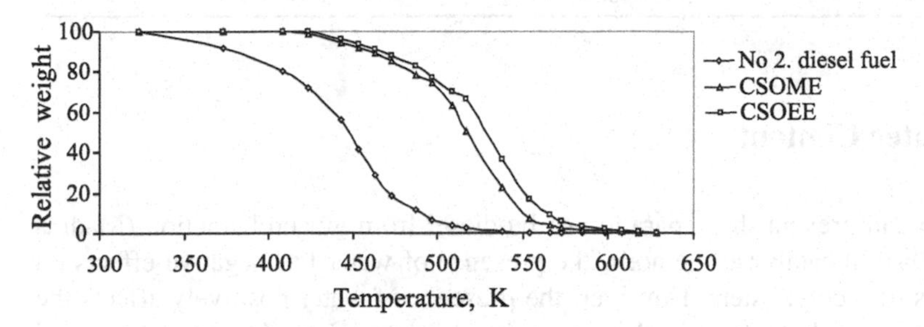

Fig. 6.4 Distillation curves of No. 2. diesel fuel and methyl ester (CSOME) and ethyl ester (CSOEE) from cottonseed oil (CSO)

6.4 Higher Combustion Efficiency of Biodiesel

The oxygen content of biodiesel improves the combustion process and decreases its oxidation potential. The structural oxygen content of a fuel improves its combustion efficiency due to an increase in the homogeneity of oxygen with the fuel during combustion. Because of this the combustion efficiency of biodiesel is higher than that of petrodiesel, and the combustion efficiency of methanol/ethanol is higher than that of gasoline. A visual inspection of the injector types would indicate no difference between biodiesel fuels and petrodiesel in testing. The overall injector coking is considerably low. Biodiesel contains 11% oxygen by weight and no sulfur. The use of biodiesel can extend the life of diesel engines because it is more lubricating than petroleum diesel fuel. Biodiesel has better lubricant properties than petrodiesel.

The higher heating values (HHVs) of biodiesels are relatively high. The HHVs of biodiesels (39 to 41 MJ/kg) is slightly lower than that of gasoline (46 MJ/kg), petrodiesel (43 MJ/kg), or petroleum (42 MJ/kg), but higher than coal (32 to 37 MJ/kg). Table 6.3 shows a comparison of chemical properties and HHVs between biodiesel and petrodiesel fuels.

Table 6.3 Comparison of chemical properties and higher heating values (HHVs) between biodiesel and D2 fuels

Chemical property	Biodiesel (methyl ester)	D2 fuel
Ash (wt.%)	0.002–0.036	0.006–0.010
Sulfur (wt.%)	0.006–0.020	0.020–0.050
Nitrogen (wt.%)	0.002–0.007	0.0001–0.003
Aromatics (vol.%)	0	28–38
Iodine number	65–156	0
HHV (MJ/kg)	39.2–40.6	45.1–45.6

6.5 Water Content

The soap can prevent the separation of biodiesel from glycerol fraction (Madras *et al.*, 2004). In catalyzed methods, the presence of water has negative effects on the yields of methyl esters. However, the presence of water positively affects the formation of methyl esters in the supercritical methanol method (Kusdiana and Saka, 2004).

6.6 Comparison of Fuel Properties and Combustion Characteristics of Methyl and Ethyl Alcohols and Their Esters

The main goals of the alcohol studies were to better understand the fuel properties of alcohol and the basic principles of conversion in order to provide a representative cross section for converting diesel engine and gasoline engine fuel into blended fuel.

Alcohols are oxygenate fuels in which the alcohol molecules have one or more oxygen molecules. The combustion heat of an alcohol decreases with increasing its oxygen content. Practically any of the organic molecules of the alcohol family can be used as a fuel. The alcohols that can be used as motor fuels are methanol (CH_3OH), ethanol (C_2H_5OH), propanol (C_3H_7OH), and butanol (C_4H_9OH). However, only two of the alcohols are technically and economically suitable as fuels for internal combustion engines (ICEs). The main production facilities of methanol and ethanol are presented in Table 6.4. Methanol is produced by a variety of processes, the most common of which is the distillation of liquid products from wood and coal, natural gas, and petroleum gas. Ethanol is produced mainly from biomass bioconversion (Bala, 2005).

In general, the physical and chemical properties and the performance of ethyl esters are comparable to those of methyl esters. Methyl and ethyl esters have almost the same heat content. The viscosities of ethyl esters are slightly higher and the cloud and pour points are slightly lower than those of methyl esters. Engine

Table 6.4 Main production facilities of methanol and ethanol

Product	Production process
Methanol	
	Distillation of liquid from wood pyrolysis
	Gaseous products from biomass gasification
	Distillation of liquid from coal pyrolysis
	Synthetic gas from biomass and coal
	Natural gas
	Petroleum gas
Ethanol	
	Fermentation of sugars and starches
	Bioconversion of cellulosic biomass
	Hydration of alkanes
	Synthesis from petroleum
	Synthesis from coal
	Enzymatic conversion of synthetic gas

tests demonstrated that methyl esters produced slightly higher power and torque than ethyl esters (Encinar *et al.*, 2002). Some desirable attributes of ethyl esters over methyl esters are their significantly lower smoke opacity, lower exhaust temperatures, and lower pour point. The ethyl esters tend to have more injector coking than the methyl esters. Some properties of fuels are given in Table 6.5.

Many alcohols have been used to make biodiesel. Issues such as cost of the alcohol, the amount of alcohol needed for the reaction, the ease of recovering and recycling the alcohol, fuel tax credits, and global warming influence the choice of alcohol. Some alcohols also require slight technical modifications to the production process such as higher operating temperatures, longer or slower mixing times, or lower mixing speeds. Since the reaction to form the esters is on a molar basis and alcohol is purchased on a volume basis, their properties make a significant difference in raw material price. It takes three moles of alcohol to react completely with one mole of triglyceride.

The systematic effect of ethyl alcohol differs from that of methyl alcohol. Ethyl alcohol is rapidly oxidized in the body to carbon dioxide and water, and, in contrast to methyl alcohol, no cumulative effect occurs. Ethanol is also a preferred alcohol in the transesterification process compared to methanol because it is derived

Table 6.5 Some properties of fuels

Fuel property	Gasoline	D2	Isoctane	Methanol	Ethanol
Cetane number	–	50	–	5	8
Octane number	96	–	100	112	107
Autoignition temperature (K)	644	588	530	737	606
Latent vaporization heat (MJ/Kg)	0.35	0.22	0.26	1.18	0.91
Lower heating value (MJ/Kg)	44.0	42.6	45.0	19.9	26.7

from agricultural products and is renewable and biologically less objectionable to the environment.

Methanol use in current-technology vehicles has some distinct advantages and disadvantages. On the plus side, methanol has a higher octane rating than gasoline. Methanol has a high vaporization heat that results in lower peak flame temperatures than gasoline and lower nitrogen oxide emissions. Combustion of methanol with higher air-to-fuel equivalence ratio results in generally lower overall emissions and higher energy efficiency. However, several disadvantages must be studied and overcome before neat methanol is considered a viable alternative to gasoline. The energy density of methanol is about half that of gasoline, reducing the distance a vehicle can travel.

There are some important differences in the combustion characteristics of alcohols and hydrocarbons. Alcohols have higher flame speeds and extended flammability limits. Pure methanol is very flammable, and its flame is colorless when ignited. The alcohols mix in all proportions with water due to the polar nature of the OH group. Low volatility is indicated by a high boiling point and high flash point. The combustion of alcohol in the presence of air can be initiated by an intensive source of localized energy, such as a flame or a spark, and the mixture can also be ignited by application of energy by means of heat and pressure, such as happens in the compression stroke of a piston engine. The high latent vaporization heat of alcohols cools the air entering the combustion chamber of the engine, thereby increasing the air density and mass flow. This leads to increased volumetric efficiency and reduced compression temperatures. The oxygen content of alcohols diminishes the heating value of the fuel in comparison with hydrocarbon fuels. The heat of combustion per unit volume of alcohol is approximately half that of isooctane.

Methanol is not miscible with hydrocarbons, and separation ensues readily in the presence of small quantities of water, particularly with a reduction in temperature. On the other hand, anhydrous ethanol is completely miscible in all proportions with gasoline, although separation may be effected by the addition of water or by cooling. If water is already present, the water tolerance is higher for ethanol than for methanol and can be improved by the addition of higher alcohols such as butanol. Benzene or acetone can also be used. The wear problem is believed to be caused by formic acid attack when methanol is used or acetic acid attack when ethanol is used.

Methanol is considerably easier to recover than ethanol. Ethanol forms an azeotrope with water so it is expensive to purify ethanol during recovery. If the water is not removed, it will interfere with the reactions. Methanol recycles easier because it does not form an azeotrope. These two factors are the reason that, even though methanol is more toxic, it is the preferred alcohol for producing biodiesel. Methanol has a flash point of 283 K, while the flash point of ethanol is 281 K, so both are considered highly flammable. One should never let methanol come into contact with skin or eyes as it can be readily absorbed. Excessive exposure to methanol can cause blindness and other health effects.

Dry methanol is very corrosive to some aluminum alloys, but additional water at 1% almost completely inhibits corrosion. It must be noted that methanol with additional water at more than 2% becomes corrosive again. Ethanol always contains some acetic acid and is particularly corrosive to aluminum alloys.

Since alcohols, especially methanol, can be readily ignited by hot surfaces, preignition can occur. It must be emphasized here that preignition and knocking in alcohol engines is a much more dangerous condition than gasoline engines. Other properties, however, are favorable to the increase of power and reduction of fuel consumption. Such properties are as follows: (1) greater number of molecules or products than in reactants, (2) extended limits of flammability, (3) high octane number, (4) high latent vaporization heat, (5) constant boiling temperature, and (6) high density.

Figures 6.5 and 6.6 show the plots of ester yield as treated with subcritical (at 503 K) and supercritical (at 523 K) methanol and ethanol as a function of reaction time. It is evident that at a subcritical state of alcohol, the reaction rate is so low and gradually increases as either pressure or temperature rises (Madras *et al.*, 2004). It was observed that increasing the reaction temperature, especially to

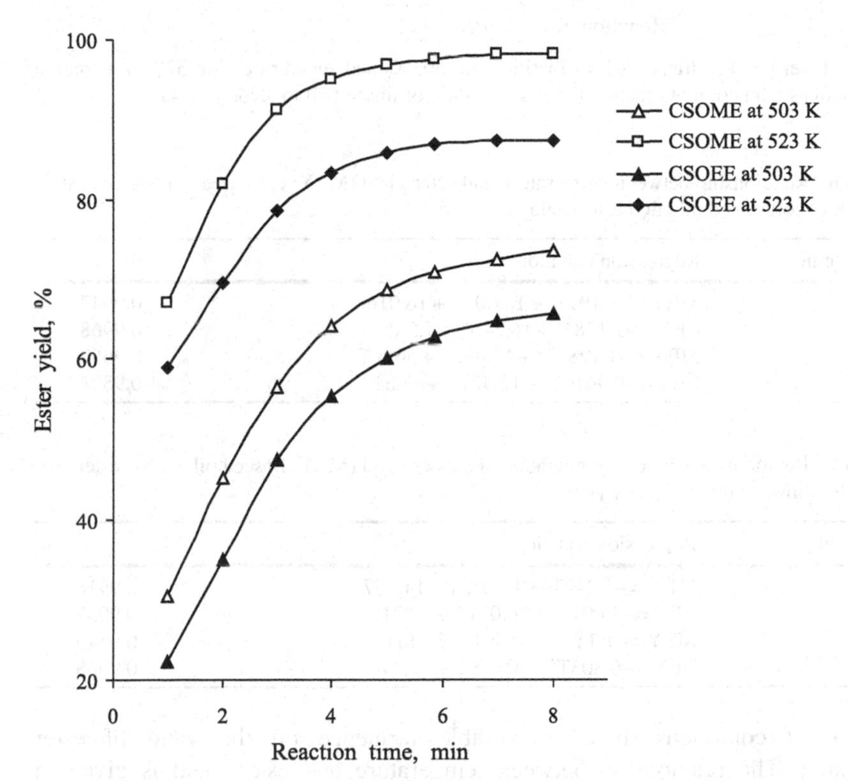

Fig. 6.5 Ester yield as treated with subcritical and supercritical methanol and ethanol as a function of reaction time (CSOME: cottonseed oil methyl ester and CSOEE: cottonseed oil ethyl ester). Molar ratio of cottonseed oil (CSO) to alcohol: 1:41

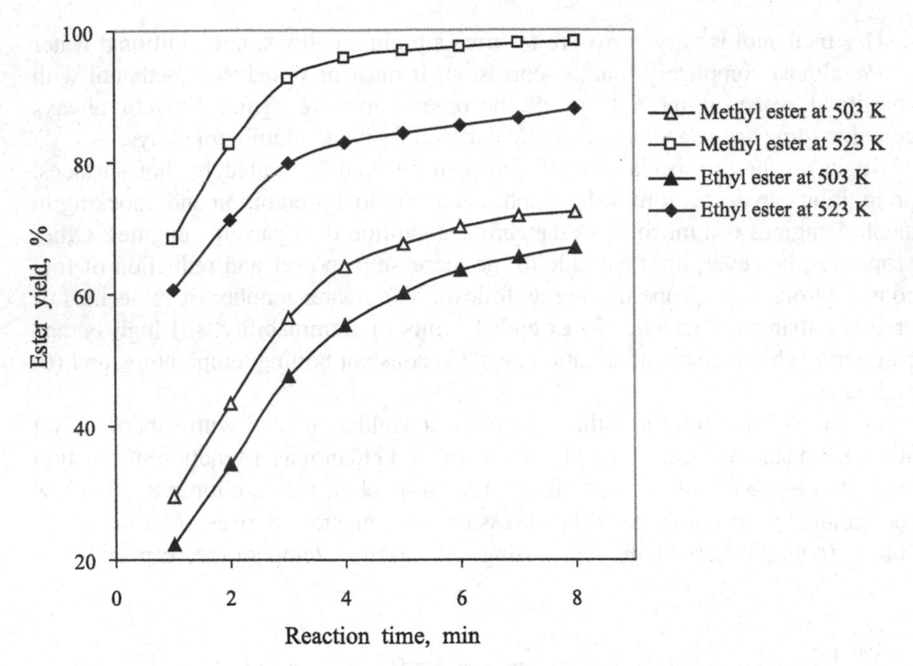

Fig. 6.6 Ester yield as treated with subcritical (at 503 K) and supercritical (at 523 K) methanol and ethanol as a function of reaction time. Molar ratio of linseed oil to alcohol: 1:41

Table 6.6 Relationship between temperature and ester yield (MEY: cottonseed oil methyl ester yield; EEY: cottonseed oil ethyl ester yield)

Temperature	Regression equation	R^2
503 K	$MEY = -1.197T^2 + 16.602T + 16.016$	0.9947
	$EEY = -1.138T^2 + 16.246T + 7.581$	0.9968
523 K	$MEY = -1.1281T^2 + 13.968T + 56.633$	0.9659
	$EEY = -0.9616T^2 + 12.421T + 48.314$	0.9898

Table 6.7 Relationship between temperature and ester yield (MEY: linseed oil methyl ester yield; EEY: linseed oil ethyl ester yield)

Temperature	Regression equation	R^2
503 K	$MEY = -1.249T^2 + 17.042T + 14.457$	0.9918
	$EEY = -1.091T^2 + 16.021T + 7.591$	0.9957
523 K	$MEY = -1.1T^2 + 13.519T + 58.464$	0.9545
	$EEY = -0.803T^2 + 10.659T + 52.348$	0.9668

supercritical conditions, had a favorable influence on the yield of ester conversion. The relationship between temperature and ester yield is given in Tables 6.6 and 6.7. The equations given in Tables 6.6 and 6.7 were obtained from Fig.s 6.5 and 6.6.

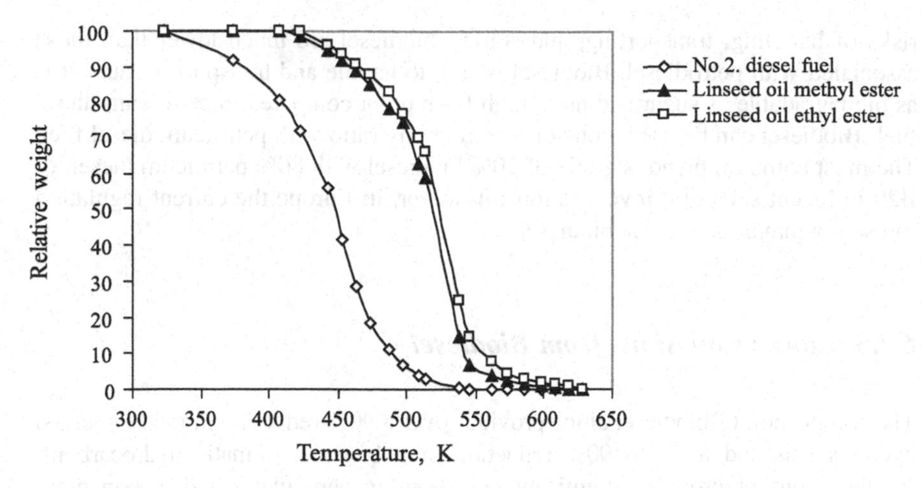

Fig. 6.7 Distillation curves of D2 fuel and linseed oil methyl and ethyl esters

Figure 6.7 shows the distillation curves of D2 fuel and linseed oil methyl and ethyl esters. As seen in Fig. 6.5, the most volatile fuel was D2 fuel. The volatility of methyl esters was higher than that of ethyl ester at all temperatures.

6.7 Advantages and Disadvantages of Biodiesels

6.7.1 Advantages of Biodiesel as Diesel Fuel

The advantages of biodiesel as a diesel fuel are its portability, ready availability, renewability, higher combustion efficiency, lower sulfur and aromatic content (Ma and Hanna, 1999; Knothe *et al.*, 2006), higher cetane number, and higher biodegradability (Mudge and Pereira, 1999; Speidal *et al.*, 2000 ; Zhang *et al.*, 2003). The main advantages of biodiesel given in the literature include its domestic origin, its potential for reducing a given economy's dependency on imported petroleum, biodegradability, high flash point, and inherent lubricity in the neat form (Mittelbach and Remschmidt, 2004; Knothe *et al.*, 2005).

6.7.2 Availability and Renewability of Biodiesel

Biodiesel is the only alternative fuel in which low-concentration biodiesel-diesel blends run on conventional unmodified engines. It can be stored anywhere that petroleum diesel fuel is stored. Biodiesel can be made from domestically produced, renewable oilseed crops such as soybean, rapeseed, and sunflower. The risks of handling, transporting, and storing biodiesel are much lower than those associated with petrodiesel. Biodiesel is safe to handle and transport because it is as biodegradable as sugar and has a high flash point compared to petroleum diesel fuel. Biodiesel can be used alone or mixed in any ratio with petroleum diesel fuel. The most common blend is a mix of 20% biodiesel with 80% petroleum diesel, or B20 in recent scientific investigations; however, in Europe the current regulation foresees a maximum 5.75% biodiesel.

6.7.3 Lower Emissions from Biodiesel

The combustion of biodiesel alone provides over a 90% reduction in total unburned hydrocarbons and a 75 to 90% reduction in polycyclic aromatic hydrocarbons. Biodiesel further provides significant reductions in particulates and carbon monoxide than petroleum diesel fuel. Biodiesel provides a slight increase or decrease in nitrogen oxides depending on the engine family and testing procedures.

Many studies on the performances and emissions of compression ignition engines, fuelled with pure biodiesel and blends with diesel oil, have been conducted and are reported in the literature (Laforgia *et al.*, 1994; Cardone *et al.*, 1998).

Fuel characterization data show some similarities and differences between biodiesel and petrodiesel fuels (Shay, 1993). The sulfur content of petrodiesel is 20 to 50 times that of biodiesels. Several municipalities are considering mandating the use of low levels of biodiesel in diesel fuel on the basis of several studies that have found hydrocarbon (HC) and particulate matter (PM) benefits from the use of biodiesel.

The use of biodiesel to reduce N_2O is attractive for several reasons. Biodiesel contains little nitrogen, as compared with petrodiesel, which is also used as a reburning fuel. The N_2O reduction is strongly dependent upon initial N_2O concentrations and only slightly dependent upon temperature, where increased temperature increases N_2O reduction. This results in lower N_2O production from fuel nitrogen species for biodiesel. In addition, biodiesel contains trace amounts of sulfur, so SO_2 emissions are reduced in direct proportion to the petrodiesel replacement.

Biodiesel has demonstrated a number of promising characteristics, including reduction of exhaust emissions (Dunn, 2001). Vegetable oil fuels have not been accepted because they are more expensive than petroleum fuels. With recent increases in petroleum prices and uncertainties concerning petroleum availability, there is renewed interest in vegetable oil fuels for compression ignition engines (CIEs) or diesel engines. Alternative fuels for CIEs have become increasingly important due to increased environmental concerns and for various several socioeconomic reasons. In this sense, vegetable oils and animal fats represent a promising alternative to conventional diesel fuel (Dorado *et al.*, 2003).

One of the most common blends of biodiesel contains 20 volume percent biodiesel and 80 volume percent conventional diesel. For soybean-based biodiesel at this concentration, the estimated emission impacts for percent change in emissions of NO_x, PM, HC, and CO were +20%, –10.1%, –21.1%, and –11.0%, respectively (EPA, 2002). The use of blends of biodiesel and diesel oil are preferred in engines in order to avoid some problems related to the decrease of power and torque and to the increase of NO_x emissions (a contributing factor in the localized formation of smog and ozone) that occurs with an increase in the content of pure biodiesel in a blend (Schumacher *et al.*, 1996). Emissions of all pollutants except NO_x appear to decrease when biodiesel is used. Figure 6.8 shows the average emission impacts of vegetable-oil-based biodiesel for CIEs. Figure 6.9 shows the average emission impacts of animal-based biodiesel for CIEs.

The use of biodiesel in a conventional diesel engine dramatically reduces the emissions of unburned hydrocarbons, carbon dioxide, carbon monoxide, sulfates, polycyclic aromatic hydrocarbons, nitrated polycyclic aromatic hydrocarbons,

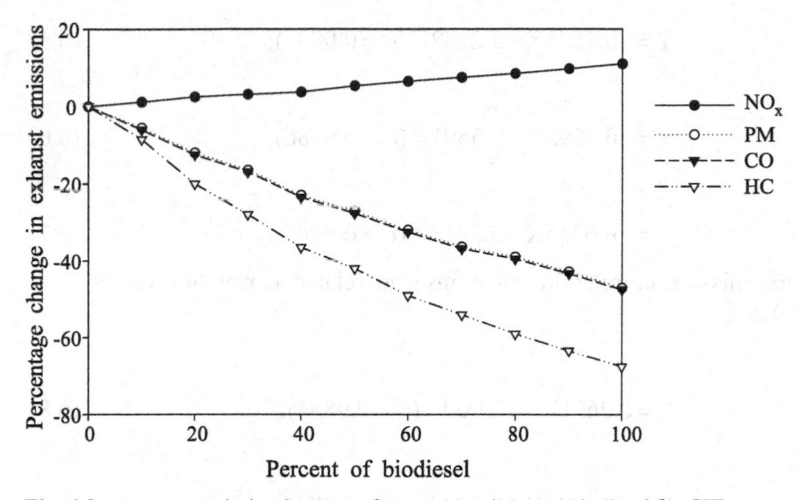

Fig. 6.8 Average emission impacts of vegetable-oil-based biodiesel for CIEs

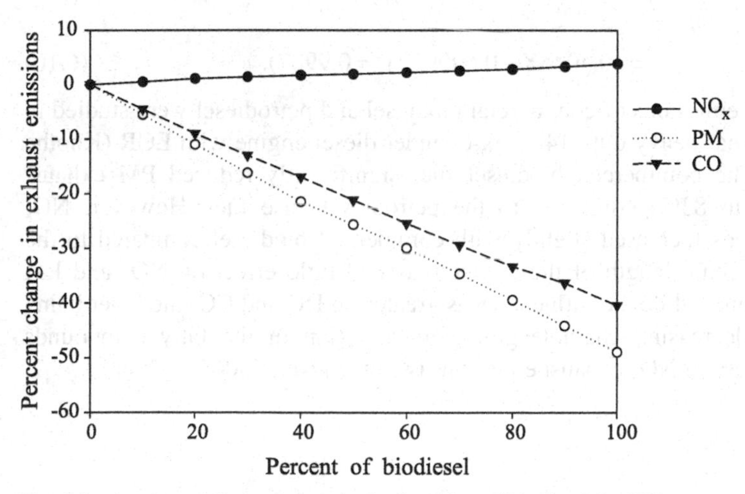

Fig. 6.9 Average emission impacts of animal-based biodiesel for CIEs

ozone-forming hydrocarbons, and particulate matter. The net contribution of carbon dioxide from biomass combustion is small (Carraretto *et al.*, 2004). Reductions in net carbon dioxide emissions are estimated at 77 to 104 g/MJ of diesel displaced by biodiesel (Tan *et al.*, 2004). These reductions increase as the amount of biodiesel blended into the diesel fuel increases. The greatest emissions reductions are seen with biodiesel.

The basic emission correlation equations and related correlation coefficients from Fig. 6.8 are as follows:

For NO_x

$$Y = 0.1078X - 0.0272 \quad (r^2 = 0.9954); \tag{6.4}$$

For PM

$$Y = -0.4673X - 2.2182 \quad (r^2 = 0.9894); \tag{6.5}$$

For CO

$$Y = -0.4695X - 2.5591 \quad (r^2 = 0.9886); \tag{6.6}$$

For HC

$$Y = -0.6715X - 5.3273 \quad (r^2 = 0.9761). \tag{6.7}$$

The basic emission correlation equations and related correlation coefficients from Fig. 6.9 are:

For NO_x

$$Y = 0.2691X + 0.2000 \quad (r^2 = 0.9854); \tag{6.8}$$

For PM

$$Y = -0.4825X - 1.0818 \quad (r^2 = 0.9986); \tag{6.9}$$

For CO

$$Y = -0.4044X - 0.6909 \quad (r^2 = 0.9987). \tag{6.10}$$

The exhaust emissions of commercial biodiesel and petrodiesel were studied in a 2003 model year heavy-duty 14 L six-cylinder diesel engine with EGR (Knothe *et al.*, 2006). The commercial biodiesel fuel significantly reduced PM exhaust emissions (75 to 83%) compared to the petrodiesel base fuel. However, NO_x exhaust emissions increased slightly with commercial biodiesel compared to the base fuel. The chain length of the compounds had little effect on NO_x and PM exhaust emissions, while the influence was greater on HC and CO, the latter being reduced with decreasing chain length. Non-saturation in the fatty compounds causes an increase in NO_x exhaust emissions (Knothe *et al.*, 2006).

6.7.4 Biodegradability of Biodiesel

Biodiesel fuels can be used as a renewable energy source to replace conventional petroleum diesel in CIEs. When degradation is caused by biological activity, especially by enzymatic action, it is called biodegradation. Biodegradability of biodiesel has been proposed as a solution for the waste problem. Biodegradable fuels such as biodiesels have an expanding range of potential applications and they are environmentally friendly. Therefore, there is growing interest in degradable diesel fuels that degrade more rapidly than conventional disposable fuels.

In recent years biodiesel has become more attractive because of its environmental benefits and the fact that it is made from renewable resources (Ma and Hanna, 1999).

Biodiesel is non-toxic and degrades about four times faster than petrodiesel. Its oxygen content improves the biodegradation process, leading to a decreased level of quick biodegradation. In comparison with petrodiesel, biodiesel shows better emission parameters. It improves the environmental performance of road transport and reduces greenhouse emissions (mainly of carbon dioxide).

As biodiesel fuels are becoming commercialized, their existence in the environment is an area of concern since petroleum oil spills constitute a major source of contamination of the ecosystem (Peterson *et al.*, 1995). Among these concerns, water quality is one of the most important issues for living systems. It is important to examine the biodegradability of biodiesel fuels and their biodegradation rates in natural waterways in case they enter the aquatic environment in the course of their use or disposal. Chemicals from biodegradation of biodiesel can be released into the environment. With the increasing interest in biodiesel, the health and safety aspects are of utmost importance, including determination of their environmental impacts in transport, storage, or processing (Ma and Hanna, 1999).

Biodegradation is degradation caused by biological activity, particularly by enzyme action, leading to significant changes in a material's chemical structure. There are many methods of biodegradation. Among them, the carbon dioxide (CO_2) evolution method is relatively simple, economical, and environmentally safe. Another method is to measure the biochemical oxygen demand (BOD) with a respirometer (Piskorz and Radlein, 1999).

The biodegradabilities of several biodiesels in the aquatic environment show that all biodiesel fuels are readily biodegradable. In one study, after 28 d all biodiesel fuels were 77 to 89% biodegraded; diesel fuel was only 18% biodegraded (Zhang, 1996). The enzymes responsible for the dehydrogenation/oxidation reactions that occur in the process of degradation recognize oxygen atoms and attack them immediately (Zhang *et al.*, 1998).

The biodegradability data of petroleum and biofuels available in the literature are presented in Table 6.8. In 28-d laboratory studies, heavy fuel oil had a low biodegradation of 11% due to its higher proportion of high-molecular-weight aromatics (Mulkins-Phillips and Stewart, 1974; Walker *et al.*, 1976). Gasoline is highly biodegradable (28%) after 28 d. Vegetables oils and their derived methyl

Table 6.8 Biodegradability data of petroleum and biofuels

Fuel sample	Degradation in 28 d (%)	Reference
Gasoline (91 octane)	28	Speidel *et al.*, 2000
Heavy fuel (Bunker C oil)	11	Mulkins-Phillips and Stewart, 1974; Walker *et al.*, 1976
Refined rapeseed oil	78	Zhang *et al.*, 1998
Refined soybeen oil	76	Zhang *et al.*, 1998
Rapeseed oil methyl ester	88	Zhang *et al.*, 1998
Sunflower seed oil methyl ester	90	Zhang *et al.*, 1998

esters (biodiesels) are rapidly degraded to reach a biodegradation rate of between 76 and 90% (Zhang *et al.*, 1998; Mudge and Pereira, 1999). In their studies Zhang *et al.* (1998) have shown that vegetable oils are slightly less degraded than their modified methyl ester.

6.7.5 Thermal Degradation of Fatty Acids During Biodiesel Production

Many vegetable oils contain polyunsaturated fatty acid chains that are methylene-interrupted rather than conjugated. The double bond of unsaturated fatty acids restricts rotation of the hydrogen atoms attached to them. Therefore an unsaturated fatty acid with a double bond can exist in two forms: the *cis* form, in which the two hydrogens are on the same "side", and the *trans* form, in which the hydrogen atoms are on opposite sides.

Trans unsaturated fatty acids, or *trans* fats, are solid fats produced artificially by heating liquid vegetable oils in the presence of metal catalysts and hydrogen. This process, partial hydrogenation, causes carbon atoms to bond in a straight configuration and remain in a solid state at room temperature (Katan *et al.*, 1995).

Physical properties that are sensitive to the effects of fatty oil oxidation include viscosity, refractive index, and dielectric constant. Figure 6.10 shows the mechanism of peroxy radical formation on a methylene group. In oxidative instability, the methylene group ($-CH_2-$) carbons between the olefinic carbons are the sites of first attack (Williard *et al.*, 1998).

Oxidation to CO_2 of biodiesel results in the formation of hydroperoxides. The formation of a hydroperoxide follows a well-known peroxidation chain mechanism. Oxidative lipid modifications occur through lipid peroxidation mechanisms in which free radicals and reactive oxygen species abstract a methylene hydrogen atom from polyunsaturated fatty acids, producing a carbon-centered lipid radical. Spontaneous rearrangement of the 1,4-pentadiene yields a conjugated diene, which reacts with molecular oxygen to form a lipid peroxyl radical. Abstraction of a proton from neighboring polyunsaturated fatty acids produces a lipid hydroperoxide (LOOH)

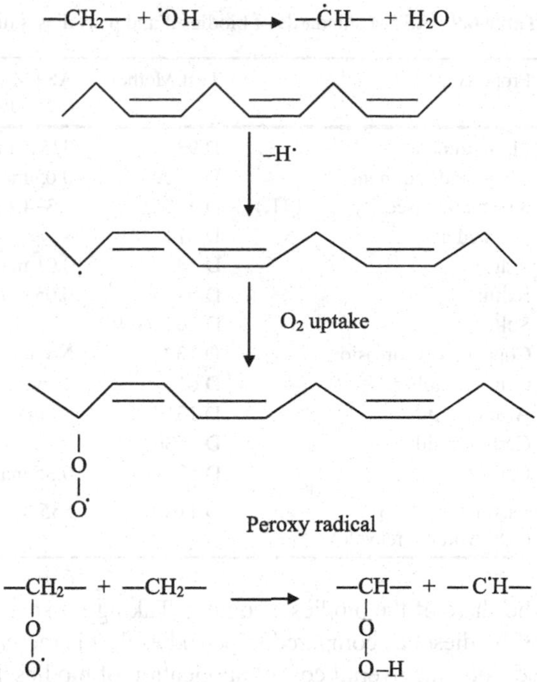

Fig. 6.10 Mechanism of peroxy radical formation on methylene group

—CH₂— + ˙OH ⟶ —ĊH— + H₂O

–H˙

O₂ uptake

Peroxy radical

and regeneration of a carbon-centered lipid radical, thereby propagating the radical reaction (Browne and Armstrong, 2000). After hydrogen is removed from such carbons oxygen rapidly attacks and a LOOH is ultimately formed where the polyunsaturation has been isomerized to include a conjugated diene. This reaction is a chain mechanism that can proceed rapidly once an initial induction period has occurred. The greater the level of unsaturation in a fatty oil or ester, the more susceptible it will be to oxidation. Once the LOOHs have formed, they decompose and interreact to form numerous secondary oxidation products including higher-molecular-weight oligomers often called polymers.

6.7.6 Disadvantages of Biodiesel as Diesel Fuel

The major disadvantages of biodiesel are its higher viscosity, lower energy content, higher cloud point and pour point, higher nitrogen oxide (NO_x) emissions, lower engine speed and power, injector coking, engine compatibility, high price, and higher engine wear.

Table 6.9 shows the fuel ASTM standards of biodiesel and petroleum diesel fuels. Important operating disadvantages of biodiesel in comparison with petrodiesel are cold start problems, lower energy content, higher copper strip corrosion, and fuel pumping difficulty from higher viscosity. This increases fuel consumption when biodiesel is used instead of pure petrodiesel, in proportion to

Table 6.9 ASTM standards of biodiesel and petrodiesel fuels

Property	Test Method	ASTM D975 (petrodiesel)	ASTM D6751 (biodiesel, B100)
Flash point	D 93	325 K min	403 K
Water and sediment	D 2709	0.05 max %vol	0.05 max %vol
Kinematic viscosity (at 313 K)	D 445	1.3–4.1 mm^2/s	1.9–6.0 mm^2/s
Sulfated ash	D 874	–	0.02 max %wt
Ash	D 482	0.01 max %wt	–
Sulfur	D 5453	0.05 max %wt	–
Sulfur	D 2622/129	–	0.05 max %wt
Copper strip corrosion	D 130	No 3 max	No 3 max
Cetane number	D 613	40 min	47 min
Aromaticity	D 1319	35 max %vol	–
Carbon residue	D 4530	–	0.05 max %mass
Carbon residue	D 524	0.35 max %mass	–
Distillation temp (90% volume recycle)	D 1160	555 K min–611 K max	–

the share of the biodiesel content. Taking into account the higher production value of biodiesel as compared to petrodiesel, this increase in fuel consumption raises in addition the overall cost of application of biodiesel as an alternative to petrodiesel.

Biodiesel has a higher cloud point and pour point compared to conventional diesel (Prakash, 1998). Neat biodiesel and biodiesel blends increase nitrogen oxide (NO_x) emissions compared with petroleum-based diesel fuel used in an unmodified diesel engine (EPA, 2002). Peak torque is lower for biodiesel than petroleum diesel but occurs at lower engine speed and generally the torque curves are flatter. Biodiesels on average decrease power by 5% compared to diesel at rated loads (Demirbas, 2006b).

References

Ali, Y., Hanna, M.A., Cuppett, S. L. 1995. Fuel properties of tallow and soybean oil esters. JAOCS 72:1557–1564.

Bala, B.K. 2005. Studies on biodiesels from transformation of vegetable oils for diesel engines. Energy Edu Sci Technol 15:1–43.

Browne, R.W., Armstrong, D. 2000. HPLC Analysis of lipid-derived polyunsaturated fatty acid peroxidation products in oxidatively modified human plasma. Clinical Chem 46:829–836.

Cardone, M., Prati, M.V., Rocco, V., Senatore, A. 1998. Experimental analysis of performances and emissions of a diesel engines fuelled with biodiesel and diesel oil blends. Proceedings of MIS–MAC V, Roma, p. 211–25. [in Italian].

Carraretto, C., Macor, A., Mirandola, A., Stoppato, A., Tonon, S. 2004. Biodiesel as alternative fuel: experimental analysis and energetic evaluations. Energy 29:2195–2211.

Demirbas, A. 2003. Biodiesel fuels from vegetable oils via catalytic and non-catalytic supercritical alcohol transesterifications and other methods: a survey. Energy Convers Mgmt 44:2093–2109.

Demirbas, A. 2006. Global biofuel strategies. Energy Edu Sci Technol 17:27–63.
Dorado, M.P., Ballesteros, E.A., Arnal, J.M., Gomez, J., Lopez, F.J. 2003. Exhaust emissions from a diesel engine fueled with transesterified waste olive oil. Fuel 82:1311–1315.
Dunn, R.O. 2001. Alternative jet fuels from vegetable-oils. Trans ASAE 44:1151–757.
Encinar, J.M., Gonzalez, J.F., Rodriguez, J.J., Tejedor, A. 2002. Biodiesel fuels from vegetable oils: Transesterification of Cynara cardunculus L. oils with ethanol. Energy Fuels 16:443–450.
EPA (US Environmental Protection Agency). 2002. A comprehensive analysis of biodiesel impacts on exhaust emissions. Draft Technical Report, EPA420-P-02-001, October 2002.
Formo, M.W. 1979. Physical properties of fats and fatty acids. Bailey's Industrial Oil and Fat Products. Vol. 1, 4th edn. Wiley, New York.
Islam, M.N., Islam, M.N., Beg, M.R.A. 2004. The fuel properties of pyrolysis liquid derived from urban solid wastes in Bangladesh. Bioresour Technol 92:181–186.
Katan, M.B., Mensink, R.P., Zock, P.L. 1995. Trans fatty acids and their effect on lipoproteins in humans. Annu Rev Nutrit 15:473–493.
Knothe, G., Krahl, J., Van Gerpen, J. (eds.) 2005. The Biodiesel Handbook. AOCS, Champaign, IL.
Knothe, G., Sharp, C.A., Ryan, T.W. 2006. Exhaust emissions of biodiesel, petrodiesel, neat methyl esters, and alkanes in a new technology engine. Energy Fuels 20:403–408.
Kusdiana, D., Saka, S. 2004. Effects of water on biodiesel fuel production by supercritical methanol treatment. Bioresour Technol 91:289–295.
Laforgia, D., Ardito, V. 1994. Biodiesel fuelled IDI engines: performances, emissions and heat release investigation. Bioresour Technol 51:53–59.
Ma, F., Hanna, M.A. 1999. Biodiesel production: a review. Bioresour Technol 70:1–15.
Madras, G., Kolluru, C., Kumar, R. 2004. Synthesis of biodiesel in supercritical fluids. Fuel 83:2029–2033.
Mittelbach, M., Remschmidt, C. 2004. Biodiesels–The Comprehensive Handbook. Karl-Franzens University Press, Graz, Austria.
Mudge, S.M., Pereira, G. 1999. Stimulating the biodegradation of crude oil with biodiesel preliminary results. Spill Sci Technol Bull 5:353–355.
Mulkins-Phillips, G.J., Stewart, J.E. 1974. Effect of environmental parameters on bacterial degradation of bunker C oil, crude oils, and hydrocarbons. Appl Microbiol 28:915–922.
Peterson, C.L., Reece, D.L., Hammon, B., Thompson, J.C., Beck, S.M. 1995. Commercalization of idaho biodiesel from ethanol and waste vegetable oil. In: ASAE meeting presentation, Chicago, 18–23 June 1995. Paper No. 956738.
Pinto, A.C., Guarieiro, L.L.N., Rezende, M.J.C., Ribeiro, N.M., Torres, E.A., Lopes, W.A., Pereira, P.A.P., Andrade, J.B. 2005. Biodiesel: an overview. J Brazil Chem Soc 16:1313–1330.
Pischinger, G.M., Falcon, A.M., Siekmann, R.W., Fernandes, F.R. 1982. Methylesters of plant oils as diesels fuels, either straight or in blends. Vegetable Oil Fuels, ASAE Publication 4-82, American Society of Agricultural Engineers, St. Joseph, MI.
Piskorz, J., Radlein, D. 1999. Determination of biodegradation rates of bio-oil by respirometry. In: Bridgwater, A.V. *et al.* (eds.) Fast Pyrolysis: A Handbook, CPL Press, Kingfisher Court Newbury, UK, pp. 119–134.
Prakash, C.B. 1998. A critical review of biodiesel as a transportation fuel in Canada. A Technical Report. GCSI – Global Change Strategies International, Canada.
Schwab, A.W., Bagby, M.O., Freedman, B. 1987. Preparation and properties of diesel fuels from vegetable oils. Fuel 66:1372–1378.
Schumacher, L.G., Borgelt, S.C., Fosseen, D., Goetz, W., Hires, W.G. 1996. Heavy-duty engine exhaust emission test using methyl ester soybean oil/diesel fuel blends. Bioresour Technol 57:31–36.
Shay, E.G. 1993. Diesel fuel from vegetable oils: status and opportunities. Biomass Bioenergy 4:227–42.

Speidel, H.K., Lightner, R.L., Ahmed, I. 2000. Biodegradability of new engineered fuels compared to conventional petroleum fuels and alternative fuels in current use. Appl Biochem Biotechnol 84-86:879–897.

Srivastava, A., Prasad, R. 2000. Triglycerides-based diesel fuels. Renew Sustain Energy Rev 4:111–133.

Walker, D., Petrakis, L., Colwell, R.R. 1976. Comparison of biodegradability of crude and fuel oils. Can J Microbiol 22:598–602.

Williard, D.E., Kaduce, T.L., Harmon, S.D., Spector, A.A. 1998. Conversion of Eicosapentaenoic acid to chain-shortened omega–3 fatty acid metabolites by peroxisomal oxidation. J Lipid Res 39:978–986.

Zhang, X. 1996. Biodegradability of biodiesel in the aquatic and soil environments. Ph.D. dissertation, Department of Biological and Agricultural Engineering, University of Idaho, Moscow, ID.

Zhang, Y., Dub, M.A., McLean, D.D., Kates, M. 2003. Biodiesel production from waste cooking oil: 2. Economic assessment and sensitivity analysis. Bioresour Technol 90:229–240.

Zhang, X., Peterson, C., Reece, D., Haws, R., Moller, G., 1998. Biodegradability of biodiesel in the aquatic environment. Trans ASAE 41: 423–430.

Chapter 7
Current Technologies in Biodiesel Production

Of the several methods available for producing biodiesel, transesterification of natural oils and fats is currently the method of choice. The purpose of the process is to lower the viscosity of the oil or fat. Transesterification is basically a sequential reaction. Triglycerides are first reduced to diglycerides, which are subsequently reduced to monoglycerides, which are finally reduced to fatty acid esters. The order of the reaction changes with the reaction conditions. The main factors affecting transesterification are the molar ratio of glycerides to alcohol, catalysts, reaction temperature and time, and free fatty acid and water content in oils and fats. Transesterification is extremely important for biodiesel. Biodiesel as it is defined today is obtained by transesterifying triglycerides with methanol. Methanol is the preferred alcohol for obtaining biodiesel because it is the cheapest alcohol. Base catalysts are more effective than acid catalysts and enzymes (Ma and Hanna, 1999). Methanol is made to react with triglycerides to produce methyl esters (biodiesel) and glycerol (Eq. 7.1).

$$\underset{\text{Triglyceride}}{C_3H_5(OOCR)_3} + \underset{\text{Methanol}}{3CH_3OH} \rightarrow \underset{\text{Methyl ester}}{3RCOOCH_3} + \underset{\text{Glycerine}}{C_3H_5(OH)_3} \tag{7.1}$$

The production processes for biodiesel are well known. There are four basic routes to biodiesel production from oils and fats:

- Base-catalyzed transesterification
- Direct acid-catalyzed transesterification
- Conversion of the oil into its fatty acids and then into biodiesel
- Non-catalytic transesterification of oils and fats

Biodiesel produced by transesterification reactions can be alkali catalyzed, acid catalyzed, or enzyme catalyzed, but the first two types have received more attention because of the short reaction times and low cost compared with the third one (Wang *et al.*, 2007). Most of the biodiesel produced today is made with the base-catalyzed reaction for several reasons:

- It involves low temperature and pressure.
- It yields high conversion (98%) with minimal side reactions and reaction time.

- It allows a direct conversion into biodiesel with no intermediate compounds.
- It requires simple construction materials.

Figure 7.1 shows the flow diagram of base-catalyzed biodiesel processing. The base-catalyzed production of biodiesel generally occurs using the following steps:

- Mixing of alcohol and catalyst
- Transesterification reaction
- Separation
- Biodiesel washing
- Alcohol removal
- Glycerine neutralization
- Product quality

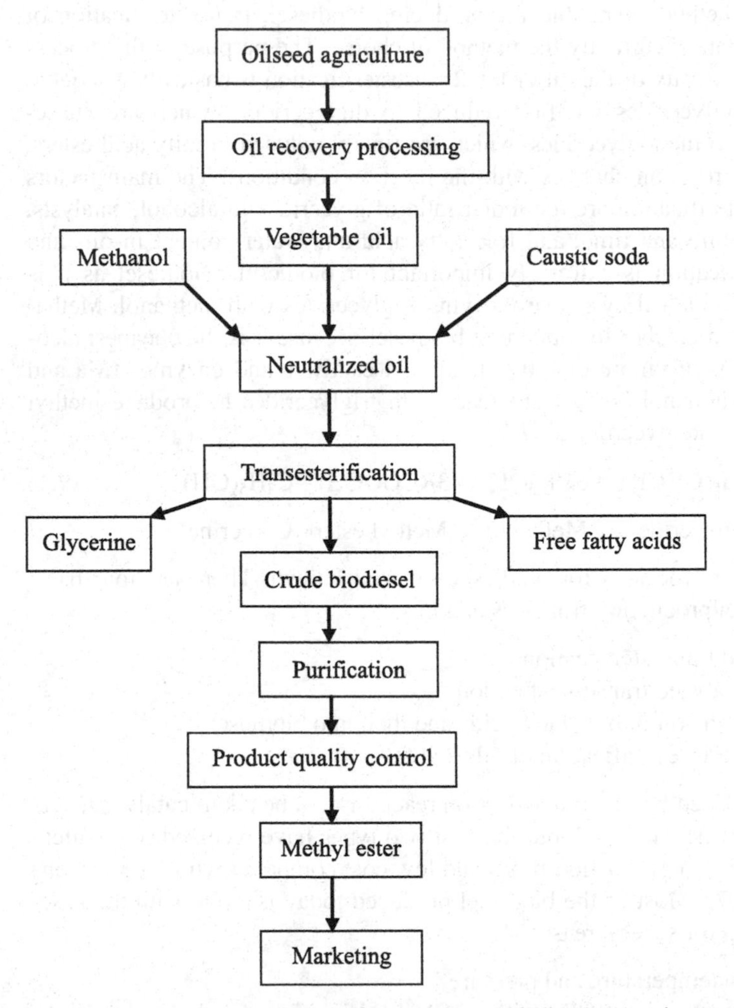

Fig. 7.1 Simplified flow diagram of base-catalyzed biodiesel processing

The basic catalyst is typically sodium hydroxide (caustic soda) or potassium hydroxide (caustic potash). It is dissolved in alcohol using a standard agitator or mixer. The methyl alcohol and catalyst mix is then charged into a closed reactor and the oil or fat is added. The reaction mix is kept just above the boiling point of the alcohol (around 344 K) to speed up the reaction, and the transesterification reaction takes place. Recommended reaction time varies from 1 to 8 h, and optimal reaction time is about 2 h (Van Gerpen *et al.*, 2004). Excess alcohol is normally used to ensure total conversion of the fat or oil into its esters. After the reaction is complete, two major products form: glycerine and biodiesel. Each has a substantial amount of the excess methanol that was used in the reaction. The reacted mixture is sometimes neutralized at this step if needed. The glycerine phase is much denser than the biodiesel phase and the two can be gravity separated with glycerine simply drawn off the bottom of the settling vessel. In some cases, a centrifuge is used to separate the two materials faster. The biodiesel product is sometimes purified by washing gently with warm water to remove residual catalyst or soaps, dried, and sent to storage (Ma and Hanna, 1999; Demirbas, 2002). For an alkali-catalyzed transesterification, the triglycerides and alcohol must be substantially anhydrous (Wright *et al.*, 1944) because water makes the reaction partially change to saponification, which produces soap. The soap lowers the yield of esters and renders the separation of ester and glycerol and the water washing difficult. Low free fatty acid content in triglycerides is required for alkali-catalyzed transesterification. If more water and free fatty acids are in the triglycerides, acid-catalyzed transesterification can be used (Keim, 1945).

When an alkali catalyst is present, the free fatty acid will react with alkali catalyst to form soap. It is common for oils and fats to contain trace amounts of water. When water is present in the reaction, it generally manifests itself through excessive soap production. The soaps of saturated fatty acids tend to solidify at ambient temperatures, so a reaction mixture with excessive soap may gel and form a semi-solid mass that is very difficult to recover. When water is present, particularly at high temperatures, it can hydrolyze the triglycerides to diglycerides and form a free fatty acid.

If an oil or fat containing a free fatty acid such as oleic acid is used to produce biodiesel, the alkali catalyst typically used to encourage the reaction will react with this acid to form soap. This reaction is undesirable because it binds the catalyst into a form that does not contribute to accelerating the reaction. Excessive soap in the products can inhibit later processing of the biodiesel, including glycerol separation and water washing. Water in the oil or fat can also be a problem (Van Gerpen *et al.*, 2004).

In some systems the biodiesel is distilled in an additional step to remove small amounts of color bodies to produce a colorless biodiesel. Once the glycerine and biodiesel phases have been separated, the excess alcohol in each phase is removed using a flash evaporation process or by distillation. The glycerine byproduct contains unused catalyst and soaps that are neutralized with sulfuric acid and sent to storage as crude glycerine. In most cases the salt is left in the glycerine. Water and alcohol are removed to produce 90% pure glycerine that is ready to be sold as crude

Table 7.1 Typical proportions for chemicals used to make biodiesel

Reactants	Amount (kg)
Fat or oil	100
Primary alcohol (methanol)	10
Catalyst (sodium hydroxide)	0.30
Neutralizer (sulfuric acid)	0.36

glycerine. Before use as a commercial fuel, the finished biodiesel must be analyzed using sophisticated analytical equipment to ensure it meets ASTM specifications.

The primary raw materials used in the production of biodiesel are vegetable oils, animal fats, and recycled greases. These materials contain triglycerides, free fatty acids, and other impurities. The primary alcohol used to form the ester is the other major feedstock. Most processes for making biodiesel use a catalyst to initiate the esterification reaction. The catalyst is required because the alcohol is sparingly soluble in the oil phase. The catalyst promotes an increase in solubility to allow the reaction to proceed at a reasonable rate. The most common catalysts used are strong mineral bases such as sodium hydroxide and potassium hydroxide. After the reaction, the base catalyst must be neutralized with a strong mineral acid.

Table 7.1 shows the typical proportions for the chemicals used to make biodiesel. The quantitative transesterification reaction of triolein obtained for methyl oleate (biodiesel) is given in Eq. (7.2):

$$\begin{array}{lllll}\text{Triolein} + 6\ \text{Methanol} & \rightarrow \text{Methyl oleate} + \text{Glycerol} + 3\ \text{Methanol} \\ 885.46\,\text{g} \quad 192.24\,\text{g} & (\text{Catalyst}) \quad 889.50\,\text{g} \quad 92.10\,\text{g} \quad 96.12\,\text{g}\end{array} \tag{7.2}$$

The most important aspects of biodiesel production to ensure trouble-free operation in diesel engines are:

- Complete transesterification reaction
- Removal of glycerine
- Removal of catalyst
- Removal of alcohol
- Removal of free fatty acids

These parameters are all specified through the biodiesel standard, ASTM D 6751. This standard identifies the parameters the pure biodiesel (B100) must meet before being used as a pure fuel or being blended with petroleum-based diesel fuel. Biodiesel, B100, specification s(ASTM D 6751 – 02 requirements) are given in Table 7.2.

The EN 14214 is an international standard that describes the minimum requirements for biodiesel produced from rapeseed fuel stock (also known as rapeseed methyl esters). Table 7.3 shows international standard (EN 14214) requirements for biodiesel.

Table 7.2 Biodiesel, B100, specifications (ASTM D 6751 – 02 requirements)

Property	Method	Limits	Units
Flash point	D 93	130 min	°C
Water and sediment	D 2709	0.050 max	% volume
Kinematic viscosity at 40°C	D 445	1.9 – 6.0	mm^2/s
Sulfated ash	D 874	0.020 max	wt.%
Total sulfur	D 5453	0.05 max	wt.%
Copper strip corrosion	D 130	No. 3 max	
Cetane number	D 613	47 min	
Cloud point	D 2500	Report	°C
Carbon residue	D 4530	0.050 max	wt.%
Acid number	D 664	0.80 max	mg KOH/g
Free glycerine	D 6584	0.020	wt.%
Total glycerine	D 6584	0.240	wt.%
Phosphorus	D 4951	0.0010	wt.%
Vacuum distillation end point	D 1160	360°C max, at 90% distilled	°C

Table 7.3 International standard (EN 14214) requirements for biodiesel

Property	Units	Lower limit	Upper limit	Test-Method
Ester content	% (m/m)	96.5	–	Pr EN 14103d
Density at 15°C	kg/m³	860	900	EN ISO 3675 / EN ISO 12185
Viscosity at 40°C	mm²/s	3.5	5.0	EN ISO 3104
Flash point	°C	>101	–	ISO CD 3679e
Sulfur content	mg/kg	–	10	–
Tar remnant (at 10% distillation remnant)	% (m/m)	–	0.3	EN ISO 10370
Cetane number	–	51.0	–	EN ISO 5165
Sulfated ash content	% (m/m)	–	0.02	ISO 3987
Water content	mg/kg	–	500	EN ISO 12937
Total contamination	mg/kg	–	24	EN 12662
Copper band corrosion (3 h at 50°C)	rating	Class 1	Class 1	EN ISO 2160
Oxidation stability at 110°C	hours	6	–	pr EN 14112k
Acid value	mg KOH/g	–	0.5	pr EN 14104
Iodine value	–	–	120	pr EN 14111
Linoleic acid methyl ester	% (m/m)	–	12	pr EN 14103d
Polyunsaturated (≥4 double bonds) methylester	% (m/m)	–	1	–
Methanol content	% (m/m)	–	0.2	pr EN 141101
Monoglyceride content	% (m/m)	–	0.8	pr EN 14105m
Diglyceride content	% (m/m)	–	0.2	pr EN 14105m
Triglyceride content	% (m/m)	–	0.2	pr EN 14105m
Free glycerine	% (m/m)	–	0.02	pr EN 14105m / pr EN 14106
Total glycerine	% (m/m)	–	0.25	pr EN 14105m
Alkali metals (Na + K)	mg/kg	–	5	pr EN 14108 / pr EN 14109
Phosphorus content	mg/kg	–	10	pr EN14107p

7.1 Biodiesel Production Processes

7.1.1 Primary Raw Materials Used in Biodiesel Production

The main factors affecting transesterification are molar ratio of glycerides to alcohol, the catalyst used, reaction temperature and pressure, reaction time, and the free fatty acid and water content in the oils.

The choice of oils or fats to be used in producing biodiesel is an important aspect of decision-making process. The cost of oils or fats directly affects the cost of biodiesel by as much as 70 to 80%. Crude vegetable oils contain some free fatty acids and phospholipids. The phospholipids are removed in a *degumming* step and the free fatty acids are removed in a *refining* step (Chapter 3, Section 3, Vegetable Oil Processing). Excess free fatty acids can be removed as soaps in a later transesterification or caustic stripping step.

7.1.1.1 Feedstock Preparation

The most desirable vegetable oils sources are soybean, canola, palm, and rape. Main animal fat sources are beef tallow, lard, poultry fat, and fish oils. Yellow greases can be mixtures of vegetable and animal sources. The free fatty acid content affects the type of biodiesel process used and the yield of fuel from that process. Other contaminants present can affect the extent of feedstock preparation necessary to use a given reaction chemistry.

From the viewpoint of a chemical reaction, refined vegetable oil is the best starting material to produce biodiesel because the conversion of pure triglyceride into fatty acid methyl ester is high, and the reaction time is relatively short. Nevertheless, waste cooking oil, if no suitable treatment is available, would be discharged and cause environmental pollution, but waste cooking oil can be collected for further purification and then biodiesel processing. This collected material is a good commercial choice to produce biodiesel due to its low cost (Zhang *et al.*, 2003; Wang *et al.*, 2007).

The most common primary alcohol used in biodiesel production is methanol, although other alcohols, such as ethanol, isopropanol, and butyl, can be used. A key quality factor for primary alcohol is the water content. Water interferes with transesterification reactions and can result in poor yields and high levels of soap, free fatty acids, and triglycerides in the final fuel.

The stoichiometric ratio for transesterification reaction requires three moles of alcohol and one mole of triglyceride to yield three moles of fatty acid ester and one mole of glycerol. Higher molar ratios result in greater ester production in a shorter time. The commonly accepted molar ratios of alcohol to vegetable oils are 3:1 to 6:1. Excess methanol (such as a 20:1 ratio) is generally necessary in batch reactors where water accumulates. Another approach is to perform the reaction in two stages: fresh methanol and sulfuric acid is reacted, removed, and replaced with fresher reactant. Much of the water is removed in the first round, and the fresh reactant in the second round drives the reaction closer to completion. The reason for using extra alcohol is that it drives the reaction closer to the 99.7% yield needed to meet the total glycerol standard for fuel-grade biodiesel. The unused alcohol must be recovered and recycled back into the process to minimize operating costs and environmental impacts (Van Gerpen *et al.*, 2004).

The most commonly used catalyst materials for converting triglycerides into biodiesel are sodium hydroxide, potassium hydroxide, and sodium methoxide. Most base-catalyst systems use vegetable oils as feedstock. The base catalysts are highly hygroscopic and form chemical water when dissolved in the alcohol reactant. They also absorb water from the air during storage.

Acid catalysts can be used for transesterification; however, they are generally considered to be too slow for industrial processing. Acid catalysts are more commonly used for the direct esterification of free fatty acids. Acid catalysts include sulfuric acid and phosphoric acid.

There is continuing interest in using lipases as enzymatic catalysts for the production of alkyl fatty acid esters. Some enzymes work on triglycerides, converting them into methyl esters, while some work on fatty acids.

Neutralizers are used to remove the base or acid catalyst from the resulting biodiesel and glycerol. If a base catalyst is used, then the neutralizer is typically an acid, and *vice versa*. If the biodiesel is being washed, the neutralizer can be added to the wash water. While hydrochloric acid is a common choice to neutralize base catalysts, as mentioned earlier, if phosphoric acid is used, the resulting salt has value as a chemical fertilizer.

7.1.2 Biodiesel Production with Batch Processing

The simplest method for producing alcohol esters is to use a batch, stirred tank reactor. Alcohol-to-triglyceride ratios from 4:1 to 20:1 (mole:mole) have been reported, with a 6:1 ratio most common. The reactor may be sealed or equipped with a reflux condenser. The operating temperature is usually about 340 K, although temperatures from 298 K to 358 K have been reported (Knothe *et al.*, 1997; Ma and Hanna, 1999; Lang *et al.*, 2001; Demirbas, 2002; Bala, 2005; Wang *et al.*, 2007). The most commonly used catalyst is sodium hydroxide, though potassium hydroxide also used. Typical catalyst loadings range from 0.3% to about 1.5%. Transesterification completion rates of 85 to 95% have been reported. Higher temperatures and higher alcohol-to-oil ratios can also enhance the completion rate. Typical reaction times range from 20 min to more than 1 h.

In transesterification the oil is first added to the system, followed by the catalyst and methanol. The system is agitated during the reaction time. Then agitation is stopped. In some processes, the reaction mixture is allowed to settle in the reactor to give an initial separation of the esters and glycerol. In other processes the

Table 7.4 Inputs and mass requirements for the Lurgi process

Input	Requirement/ton biodiesel
Feedstock	1,000 kg vegetable oil
Steam requirement	415 kg
Electricity	12 kWh
Methanol	96 kg
Catalyst	5 kg
Hydrochloric acid (37%)	10 kg
Caustic soda (50%)	1.5 kg
Nitrogen	1 Nm^3
Process water	20 kg

reaction mixture is pumped into a settling vessel or is separated using a centrifuge (Van Gerpen *et al.*, 2004). The alcohol is removed from both the glycerol and ester stream using an evaporator or a flash unit. The esters are neutralized, washed gently using warm, slightly acidic water to remove residual methanol and salts, and then dried. The finished biodiesel is then transferred to storage. The glycerol stream is neutralized and washed with soft water. The glycerol is then sent to the glycerol refining unit.

High free fatty acid feedstocks will react with the catalyst and form soaps if they are fed to a base-catalyzed system. The maximum amount of free fatty acids acceptable in a base-catalyzed system is less than 2%, and preferably less than 1%.

The Lurgi process is shown as a two-step reactor. Most of the glycerine is recovered after the first stage where a rectifying column leads to separation of the excess methanol and crude glycerine. The methyl ester output of the second stage is purified, to some extent, of residual glycerine and methanol by a wash column. Table 7.4 shows the inputs and mass requirements for the Lurgi process.

7.1.3 Biodiesel Production with Continuous Process

Conventionally, transesterification can be performed using alkaline, acid, or enzyme catalysts (Ma and Hana, 1999; Fukuda *et al.*, 2001; Zhang *et al.*, 2003). As alkali-catalyzed systems are very sensitive to both water and free fatty acid content, the glycerides and alcohol must be substantially anhydrous because water makes the reaction partially change to saponification, which produces soaps, thus consuming the catalyst and reducing the catalytic efficiency, as well as causing an increase in viscosity, formation of gels, and difficulty in separations (Ma and Hana, 1999; Fukuda *et al.*, 2001; Zhang *et al.*, 2003).

There are several processes that use intense mixing, either from pumps or motionless mixers, to initiate the esterification reaction. A popular variation of the batch process is the use of continuous stirred tank reactors in series. Instead of allowing time for the reaction in an agitated tank, the reactor is tubular. The reaction mixture moves through this type of reactor in a continuous plug, with little mixing in the axial direction. The result is a continuous system that requires rather short residence times, as low as 6 to 10 min, for near completion of the reaction.

7.1.4 Biodiesel Production with Non-catalyzed Transesterification

There are two non-catalyzed transesterification processes. These are the BIOX processBIOX process and the supercritical alcohol (methanol) process.

7.1.4.1 Biodiesel Production with BIOX Process

The BIOX process is a new Canadian process developed originally by Professor David Boocock of the University of Toronto that has attracted considerable attention. Dr. Boocock has transformed the production process through the selection of inert cosolvents that generate an oil-rich one-phase system. This reaction is over 99% complete in seconds at ambient temperatures, compared to previous processes that required several hours. BIOX is a technology development company that is a joint venture of the University of Toronto Innovations Foundation and Madison Ventures Ltd. (WS 4). Its process uses base-catalyzed transesterification (specifically transmethylation) of fatty acids to produce methyl esters. It is a continuous process and is not feedstock specific. The unique feature of the BIOX

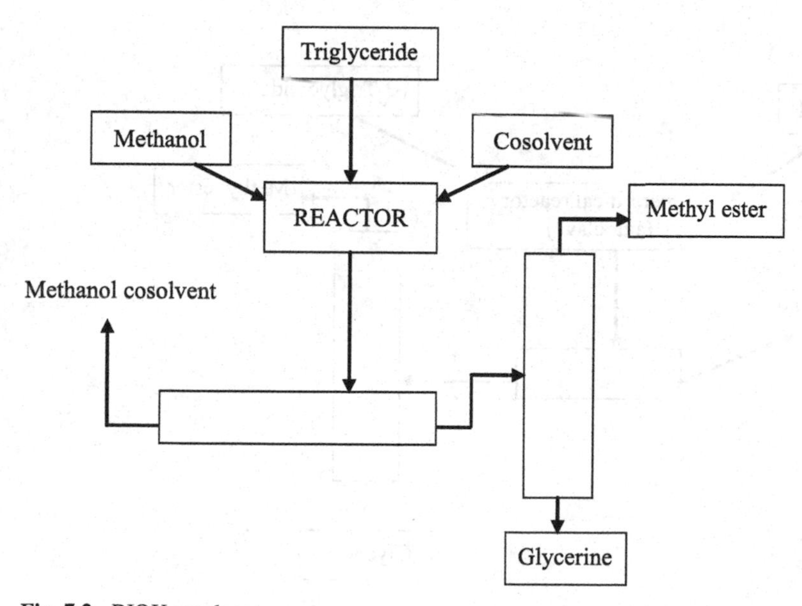

Fig. 7.2 BIOX cosolvent process

process is that it uses inert reclaimable cosolvents in a single-pass reaction taking only seconds at ambient temperature and pressure. The developers are aiming to produce biodiesel that is cost competitive with petrodiesel. The BIOX process handles not only grain-based feedstocks but also waste cooking greases and animal fats (Van Gerpen *et al.*, 2004).

The BIOX process uses a cosolvent, tetrahydrofuran, to solubilize the methanol. Cosolvent options are designed to overcome slow reaction times caused by the extremely low solubility of the alcohol in the triglyceride phase. The result is a fast reaction, on the order of 5 to 10 min, and no catalyst residues in either the ester or the glycerol phase. Figure 7.2 shows BIOX cosolvent process.

7.1.4.2 Supercritical Alcohol Process

Biodiesel fuel is one of the most promising bioenergies and can be produced from oils/fats through transesterification. A current commercial process for biodiesel production involves the use of alkali catalyst, followed by the removal of the catalyst and saponified products from free fatty acids. In addition, water-containing waste oils/fats depress the catalyst activity. These cannot allow the low-quality feedstocks, such as waste cooking oil and waste industrial oil, available for their efficient utilization. Figure 7.3 shows the supercritical transesterification process.

To overcome these problems, Saka and Kusdiana (2001) and Demirbas (2002) were the first to propose that biodiesel fuels may be prepared from vegetable oils via non-catalytic transesterification with supercritical methanol (SCM). A novel process of biodiesel fuel production has been developed by a non-catalytic super-

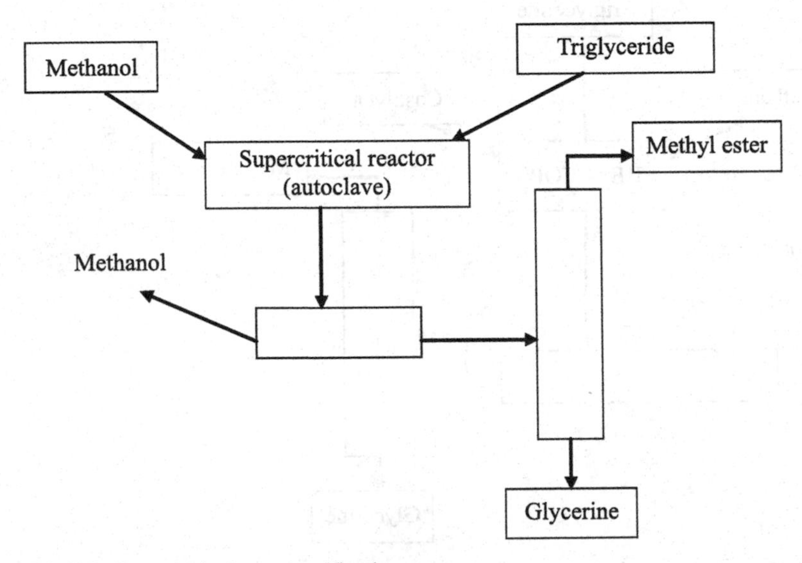

Fig. 7.3 Supercritical transesterification process

critical methanol method. Supercritical methanol is believed to solve the problems associated with the two-phase nature of normal methanol/oil mixtures by forming a single phase as a result of the lower value of the dielectric constant of methanol in the supercritical state. As a result, the reaction was found to be complete in a very short time (Han *et al.*, 2005). In contrast to catalytic processes under barometric pressure, the supercritical methanol process is non-catalytic, involves much simpler purification of products, has a lower reaction time, is more environmentally friendly, and requires lower energy use. However, the reaction requires temperatures of 525 to 675 K and pressures of 35 to 60 MPa (Demirbas, 2003; Kusdiana and Saka, 2004).

Supercritical transesterification is carried out in a high-pressure reactor (autoclave). In a typical run, the autoclave is charged with a given amount of vegetable oil and liquid methanol with changed molar ratios. The autoclave is supplied with heat from an external heater, and power is adjusted to give an approximate heating time of 15 min. The temperature of the reaction vessel can be measured with an iron-constantan thermocouple and controlled at ± 5 K for 30 min. Transesterification occurs during the heating period. After each run, the gas is vented, and the autoclave is poured into a collecting vessel. The remaining contents are removed from the autoclave by washing with methanol.

A method for non-catalytic biodiesel production by supercritical methanol treatment (>512 K, >8.09 MPa) has been under development since 1998 (Kusdiana and Saka, 2001). Variables affecting the reaction were investigated followed by a proposal that the optimum conditions were 623 K, 20 MPa, and 9 min. Compared with the alkali-catalyzed method, the supercritical methanol method has advantages in terms of reaction time and purification step.

7.1.5 Basic Plant Equipment Used in Biodiesel Production

The basic plant equipments used in biodiesel production are reactors, pumps, settling tanks, centrifuges, distillation columns, and storage tanks.

The reactor is the only place in the process where chemical conversion occurs. Reactors can be grouped into two broad categories, batch reactors and continuous reactors. In the batch reactor, the reactants are fed into the reactor at the determined amount. The reactor is then closed, and the desired reaction conditions are set. The chemical composition within the reactor changes with time. The construction materials are an important consideration for the reactor and storage tanks. For the base-catalyzed transesterification reaction, stainless steel is the preferred material for the reactor (Van Gerpen *et al.*, 2004).

Key reactor variables that dictate conversion and selectivity are temperature, pressure, reaction time (residence time), and degree of mixing. In general, increasing the reaction temperature increases the reaction rate and, hence, the conversion for a given reaction time. Increasing the temperature in the transesterification reaction does impact the operating pressure.

Two reactors within the continuous reactor category are continuous stirred tank reactors (CSTRs) and plug flow reactors (PFRs). For CSTRs, the reactants are fed into a well-mixed reactor. The composition of the product stream is identical to the composition within the reactor. Hold-up time in a CSTR is given by a residence time distribution. For PFRs, the reactants are fed into one side of the reactor. The chemical composition changes as the material moves in plug flow through the reactor (Van Gerpen *et al.*, 2004).

The pumps play the key role in moving chemicals through the manufacturing plant. The most common type of pump in the chemical industry is a centrifugal pump. The primary components of a centrifugal pump are (1) a shaft, (2) a coupling attaching the shaft to a motor, (3) bearings to support the shaft, (4) a seal around the shaft to prevent leakage, (5) an impeller, and (6) a volute, which converts the kinetic energy imparted by the impeller into the feet or head. The gear pumps are generally used in biodiesel plants. There are a number of different types of positive displacement pumps including gear pumps (external and internal) and lobe pumps. External gear pumps generally have two gears with an equal number of teeth located on the outside of the gears, whereas internal gear pumps have one larger gear with internal teeth and a smaller gear with external teeth (Van Gerpen *et al.*, 2004).

The separation of biodiesel and glycerine can be achieved using a settling tank. While a settling tank may be cheaper, a centrifuge can be used to increase the rate of separation relative to a settling tank. Centrifuges are most typically used to separate solids and liquids, but they can also be used to separate immiscible liquids of different densities. In a centrifuge separation is accomplished by exposing the mixture to a centrifugal force. The denser phase will be preferentially separated to the outer surface of the centrifuge. The choice of appropriate centrifuge type and size are predicated on the degree of separation needed in a specific system.

An important separation device for miscible fluids with similar boiling points (*e.g.*, methanol and water) is the distillation column. Separation in a distillation column is predicated on the difference in volatilities (boiling points) between chemicals in a liquid mixture. In a distillation column the concentrations of the more volatile species are enriched above the feed point and the less volatile species are enriched below the feed point.

References

Bala, B.K. 2005. Studies on biodiesels from transformation of vegetable oils for diesel engines. Energy Edu Sci Technol 15:1–43.

Demirbas, A. 2002. Biodiesel from vegetable oils via transesterification in supercritical methanol. Energy Convers Mgmt 43:2349–56.

Fukuda, H., Kondo, A., Noda, H. 2001. Biodiesel fuel production by transesterification of oils. J Biosci Bioeng 92:405–416.

Madras, G., Kolluru, C., Kumar, R. 2004. Synthesis of biodiesel in supercritical fluids. Fuel 83:2029–2033.

Han, H., Cao, W., Zhang, J. 2005. Preparation of biodiesel from soybean oil using supercritical methanol and CO_2 as co-solvent. Process Biochem 40:3148–3151.

Keim, G.I. 1945. Process for treatment of fatty glycerides. US Patent 2,383–601.

Knothe, G., Dunn, R.O., Bagby, M.O. 1997. Biodiesel: the use of vegetable oils and their derivatives as alternative diesel fuels. Am Chem Soc Symp Ser 666:172–208.

Kusdiana, D., Saka, S. 2001. Kinetics of transesterification in rapeseed oil to biodiesel fuels as treated in supercritical methanol. Fuel 80:693–698.

Kusdiana, D., Saka, S. 2004. Effects of water on biodiesel fuel production by supercritical methanol treatment. Bioresour Technol 91:289–295.

Lang, X., Dalai, A.K., Bakhshi, N.N., Reaney, M.J., Hertz, P.B. 2001. Preparation and characterization of bio-diesels from various bio-oils. Bioresour Technol 80:53–63.

Ma, F., Hanna, M.A. 1999. Biodiesel production: a review. Bioresour Technol 70:1–15.

Saka, S., Kusdiana, D. 2001. Biodiesel fuel from rapeseed oil as prepared in supercritical methanol. Fuel 80:225–231

Van Gerpen, J., Shanks, B., Pruszko, R., Clements, D., Knothe, G. 2004. Biodiesel production technology. National Renewable Energy Laboratory. 1617 Cole Boulevard, Golden, CO. Paper contract No. DE-AC36-99-GO10337.

Wang, Y., Ou, S., Liu, P., Zhang, Z. 2007. Preparation of biodiesel from waste cooking oil via two-step catalyzed process. Energy Convers Mgmt 48:184–188.

Wright, H.J., Segur, J.B., Clark, H.V., Coburn, S.K., Langdon, E.E., DuPuis, R.N. 1944. A report on ester interchange. Oil Soap 21:145–148.

Zhang, Y., Dube, M.A., McLean, D.D., Kates, M. 2003. Biodiesel production from waste cooking oil: 1. Process design and technological assessment. Bioresour Technol 89:1–16.

Zhang, Y., Dub, M.A., McLean, D.D., Kates, M. 2003. Biodiesel production from waste cooking oil: 2. Economic assessment and sensitivity analysis. Bioresour Technol 90:229–240.

Chapter 8
Engine Performance Tests

Biodiesel and petroleum-based diesel possess similar fuel properties such as viscosity, heating value, boiling temperature, cetane number, *etc.* For this reason, biodiesel may be used in standard diesel engines. The only modifications required are a two- to three-degree retardation of injection timing and replacement of all natural rubber seals with synthetic ones due to the solvent characteristics of biodiesel. The kinematic viscosity values of biodiesels are between 3.6 and 4.6 mm^2/s. The kinematic viscosity of D2 fuel is 2.7 mm^2/s at 311 K. Biodiesel is slightly heavier than fossil diesel (860 to 895 kg/m^3 for biodiesel, *ca.* 820 to 860 kg/m^3 for petroleum-based diesel at 288 K), but this fact does not prevent its mixing with petroleum-based diesel for blended application. A comparison of fuel properties of biodiesel and D2 fuels is given in Table 8.1.

From an operational point of view, biodiesel has about 90% of the energy content of petroleum diesel, measured on a volumetric basis. Due to this fact, on average the use of biodiesel reduces the fuel economy and power of an engine by about 10% in comparison with petroleum diesel. The reason for this reduction stems mainly from the oxygen content of biodiesel, the ensuing better combustion process, and improved lubricity, which partly compensate for the impact of the lower energy content. Biodiesel is an oxygenated compound. Oxygenates are just

Table 8.1 Comparison of fuel properties of biodiesel and D2 fuels

Property	Biodiesel	D2
Specific gravity, kg/m^3	860–895	840–860
Cetane number	46–70	47–55
Cloud point, K	262–289	256–265
Pour point, K	258–286	237–243
Flash point, K	408–423	325–350
Sulfur, wt.%	0.0000–0.0024	0.04–0.01
Ash, wt.%	0.002–0.01	0.06–0.01
Iodine number	60–135	–
Kinematic viscosity, at 313 K	3.6–5.0	1.9–3.8
Higher heating value, MJ/kg	39.3–39.8	45.3–46.7

preused hydrocarbons having a structure that provides a reasonable antiknock value. Also, as oxygenates contain oxygen, fuel combustion is more efficient, reducing hydrocarbons in exhaust gases. The only disadvantage is that oxygenated fuel has less energy content. For the same efficiency and power output, more fuel has to be burned. On the other hand, biodiesel blends are safer than pure petroleum diesel because biodiesel has a higher flash point. Most operational disadvantages of pure biodiesel, *e.g.*, replacement of natural rubber seals, cold start problems, *etc.*, do not arise with blended kinds of biodiesel.

Fuel characterization data show some similarities and differences between biodiesel fuels and diesel (Shay, 1993):

- Specific weight is higher for biodiesel, heat of combustion is lower, and viscosities are 1.3 to 1.6 times that of D2 fuel.
- Pour points for biodiesel fuels vary from 274 to 298 K higher for biodiesel fuels depending on the feedstock.
- Sulfur content for biodiesel fuel is 20 to 50% that of D2 fuel.
- Methyl esters have higher levels of injector coking than D2 fuel.

In cities across the globe, the personal automobile is the single greatest polluter, as emissions from millions of vehicles on the road add up to a planetwide problem. The biodiesel impacts on exhaust emissions vary depending on the type of biodiesel and on the type of conventional diesel. Blends of up to 20% biodiesel mixed with petroleum diesel fuels can be used in nearly all diesel equipment and are compatible with most storage and distribution equipment. Using biodiesel in a conventional diesel engine substantially reduces emissions of unburned hydrocarbons, carbon monoxide, sulfates, polycyclic aromatic hydrocarbons, nitrated polycyclic aromatic hydrocarbons, and particulate matter. These reductions increase as the amount of biodiesel blended into diesel fuel increases. In general, biodiesel increases NO_x emissions when used as fuel in diesel engines.

Biodiesel provides significant lubricity improvement over petroleum diesel fuel. Lubricity results of biodiesel and petroleum diesel using industry test methods indicate that there is a marked improvement in lubricity when biodiesel is added to conventional diesel fuel. Even biodiesel levels below 1% can provide up to a 30% increase in lubricity.

In general, biodiesel will soften and degrade certain types of elastomers and natural rubber compounds over time. Using high-percent blends can impact fuel system components (primarily fuel hoses and fuel pump seals) that contain elastomer compounds incompatible with biodiesel. Manufacturers recommend that natural or butyl rubbers not be allowed to come in contact with pure biodiesel. Biodiesel will lead to degradation of these materials over time, although the effect is lessened with biodiesel blends.

8.1 Engine Combustion Process and Combustion-related Concepts

A typical engine combustion reaction is as follows:

$$\text{Fuel} + \text{Air}\ (N_2 + O_2) \rightarrow CO_2 + CO + H_2O + N_2 + O_2 + (HC) + O_3 + NO_2 + SO_2. \quad (8.1)$$

Carbon dioxide (CO_2) is a colorless, odorless, non-poisonous gas that results from fossil fuel combustion and is a normal constituent of ambient air. CO_2 does not directly impair human health, but it is a "greenhouse gas" that traps the earth's heat and contributes to the potential for global warming.

Carbon monoxide (CO) is a colorless, odorless, toxic gas produced by the incomplete combustion of carbon-containing substances. One of the major air pollutants, it is emitted in large quantities by exhaust from petroleum-fuel-powered vehicles. CO is emitted directly from vehicle tailpipes. Incomplete combustion is most likely to occur at low air-to-fuel ratios in the engine. These conditions are common during vehicle starting when air supply is restricted, when cars are not tuned properly, and at altitudes where "thin" air effectively reduces the amount of oxygen available for combustion. Two thirds of CO emissions come from transportation sources, with the largest contribution coming from highway motor vehicles. In urban areas, the motor vehicle contribution to CO pollution can exceed 90%.

Under the high pressure and temperature conditions in an engine, nitrogen and oxygen atoms in the air react to form various nitrogen oxides, collectively known as NO_x. Nitrogen oxides, like hydrocarbons, are precursors to the formation of ozone. They also contribute to the formation of acid rain. Nitric oxide (NO) is the precursor of ozone, NO_2, and nitrate which are usually emitted from combustion processes. Converted into nitrogen dioxide (NO_2) in the atmosphere, it then becomes involved in the photochemical process and/or particulate formation. Nitrogen oxides (NO_x) are gases formed in great part from atmospheric nitrogen and oxygen when combustion takes place under conditions of high temperature and high pressure; they are considered major air pollutants and precursors of ozone.

Hydrocarbon (HC) emissions result when fuel molecules in the engine burn only partially. Some HC compounds are major air pollutants; they may be active participants in the photochemical process or affect health. HCs react in the presence of nitrogen oxides and sunlight to form ground-level ozone, a major component of smog. Ozone irritates the eyes, damages the lungs, and aggravates respiratory problems. It is our most widespread and intractable urban air pollution problem. A number of exhaust hydrocarbons are also toxic, with the potential to cause cancer. Hydrocarbon pollutants also escape into the air through fuel evaporation, and evaporative losses can account for a majority of the total HC pollution from current model cars on hot days when ozone levels are highest.

Sulfur oxides (SO_x) are pungent, colorless gases formed primarily by the combustion of sulfur-containing fossil fuels, especially coal and oil. Considered major air pollutants, SO_x may impact human health and damage vegetation.

Ozone (O_3) is a pungent, colorless, toxic gas. Close to the earth's surface it is produced photochemically from HCs, oxides of nitrogen, and sunlight and is a major component of smog. At very high altitudes, it protects the earth from harmful ultraviolet radiation.

Particulate matter (PM) is tiny solid or liquid particles of soot, dust, smoke, fumes, and aerosols. The size of the particles (10 microns or smaller) allows them to easily enter the air sacs in the lungs where they may be deposited, resulting in adverse health effects. PM also causes visibility reduction and is a constituent of air pollutant.

Smog is a term used to describe many air pollution problems. The word smog is a contraction of smoke and fog. Soot is made up of very fine carbon particles that appear black when visible.

Combustion is a basic chemical process that releases energy from a fuel and air mixture. For combustion to occur, fuel, oxygen, and heat must be present together. Combustion is the chemical reaction of a particular substance with oxygen. Combustion represents a chemical reaction during which from certain matters other simple matters are produced, this is a combination of inflammable matter with oxygen of the air accompanied by heat release. The quantity of heat evolved when one mole of a hydrocarbon is burned to CO_2 and water is called the heat of combustion. Combustion to CO_2 and water is characteristic of organic compounds; under special conditions it is used to determine their carbon and hydrogen content (Demirbas, 2006a). During combustion the combustible part of fuel is subdivided into a volatile part and a solid residue. During heating it evaporates together with a part of carbon in the form of hydrocarbons, combustible gases, and CO released by thermal degradation of the fuel. CO is mainly formed by the following reactions: (a) from reduction of CO_2 with unreacted C,

$$CO_2 + C \rightarrow 2CO, \tag{8.2}$$

and (b) from the degradation of carbonyl fragments (-CO) in the fuel molecules at 600 to 750 K.

The combustion process is started by heating the fuel above its ignition temperature in the presence of oxygen or air. Under the influence of heat, the chemical bonds of the fuel are cleaved. If complete combustion occurs, the combustible elements (C, H, and S) react with the oxygen content of the air to form CO_2, H_2O, and mainly SO_2.

If not enough oxygen is present or the fuel and air mixture is insufficient, then the burning gases are partially cooled below the ignition temperature and the combustion process stays incomplete. The flue gases then still contain combustible components, mainly carbon monoxide (CO), unburned carbon (C), and various hydrocarbons (C_xH_y).

The standard measure of the energy content of a fuel is its heating value (HV), sometimes called the calorific value or heat of combustion. In fact, there are multiple values for the HV, depending on whether it measures the enthalpy of combustion (ΔH) or the internal energy of combustion (ΔU), and whether for a fuel containing hydrogen product water is accounted for in the vapor phase or the condensed (liquid) phase. With water in the vapor phase, the lower heating value (LHV) at constant pressure measures the enthalpy change due to combustion (Jenkins *et al.*, 1998). The HV is obtained by the complete combustion of a unit quantity of solid fuel in an oxygen-bomb calorimeter under carefully defined conditions. The gross heat of combustion or higher heating value (GHC or HHV) is obtained by the oxygen-bomb calorimeter method as the latent heat of moisture in the combustion products is recovered.

8.2 Engine Performance Tests

The methyl ester of vegetable oil was evaluated as a fuel in compression ignition engines (CIEs) (Isigigur *et al.*, 1994; Kusdiana and Saka, 2001). The authors concluded that the performance of the esters of vegetable oil did not differ greatly from that of diesel fuel. The brake power of biodiesel was nearly the same as with diesel fuel, while the specific fuel consumption was higher than that of D2. Based on crankcase oil analysis, engine wear rates were low, but some oil dilution did occur. Carbon deposits inside the engine were normal, with the exception of intake valve deposits. The results showed the transesterification treatment decreased the injector coking to a level significantly lower than that observed with D2 fuel (Demirbas, 2003). Although most researchers agree that vegetable oil ester fuels are suitable for use in CIEs, a few contrary results have also been obtained. The results of these studies point out that most vegetable oil esters are suitable as diesel substitutes but that more long-term studies are necessary for commercial utilization to become practical.

Blends of up to 20% biodiesel mixed with petroleum diesel fuels can be used in nearly all diesel equipment and are compatible with most storage and distribution equipment. Higher blends, even B100, can be used in many engines built with little or no modification. Transportation and storage, however, require special management. Material compatibility and warranty issues have not been resolved with higher blends.

8.2.1 Alcohol-diesel Emulsions

Because alcohols have limited solubility in diesel fuel, a stable emulsion must be formed that will allow it to be injected before separation occurs. A hydro-shear emulsification unit can be used to produce emulsions of diesel alcohol. However, the emulsion can only remain stable for 45 s. In addition, 12% alcohol (energy basis) is the maximum percentage. This kind of method has several problems: (a) high specific fuel consumption at low speeds, (b) high cost, and (c) instability. Therefore, other methods have been developed.

8.2.2 *Using Microemulsions for Vegetable Oil*

To reduce the high viscosity of vegetable oils, microemulsions with immiscible liquids such as methanol and ethanol and ionic or non-ionic amphiphiles have been studied (Ma and Hanna, 1999; Schwab *et al.*, 1987). Short engine performances of both ionic and non-ionic microemulsions of ethanol in soybean oil were nearly as good as that of D2 fuel (Goering *et al.*, 1982).

All microemulsions with butanol, hexanol, and octanol met the maximum viscosity requirement for D2 fuel. The 2-octanol was found to be an effective amphiphile in the micellar solubilization of methanol in triolein and soybean oil (Ma and Hanna, 1999; Schwab *et al.*, 1987).

8.2.3 *Diesel Engine Fumigation*

A fumigation system injects a gaseous or liquid fuel into the intake air stream of a CIE. This fuel burns and becomes a partial contributor to the power-producing fuel. While alcohol and gasoline may be used, gaseous fumigation seems to exhibit the best overall power yields, performance, and emissions benefits. Rudolph Diesel's 1901 patent mentions the diesel fumigation process. Not until the 1940s were there any commercial fumigation applications. Fumigation with propane was studied as a means to reduce injector coking (Peterson *et al.*, 1987).

Fumigation is a process of introducing alcohol into a diesel engine (up to 50%) by means of a carburetor in the inlet manifold. At the same time, the diesel pump operates at a reduced flow. In this process, D2 fuel is used for generating a pilot flame, and alcohol is used as a fumigated fuel (Demirbas, 2005).

Alcohol is used as a fumigated fuel. At low loads, the quantity of alcohol must be reduced to prevent misfire. On the other hand, at high loads, the quantity of alcohol must also be reduced to prevent preignition.

8.2.4 *Dual Injection*

In dual injection systems, a small amount of diesel is injected as a pilot fuel for the ignition source. A large amount of alcohol is injected as the main fuel. It must be noted that the pilot fuel must be injected prior to injection of alcohol. Some ideal results can be achieved when this method is used. The thermal efficiency is better. At the same time, NO_x emissions are lower. Moreover, CO emissions and HC emissions are the same. However, the system requires two fuel pumps, thus resulting in a high cost. Meanwhile, alcohol requires additives for lubricity (Demirbas, 2006b).

8.2.5 *Injector Coking*

A visual inspection of the injector types would indicate no difference between the biodiesels when tested on D2 fuel. The overall injector coking is very low. Linear regression can be used to compare injector coking, viscosity, percent of biodiesel, total glycerol, and combustion heat data with the other physical properties such as flash point and cetane number (Demirbas, 2005).

8.2.6 *Heated Surfaces*

Alcohol can ignite when in contact with hot surfaces. For this reason, glow plugs can be used as a source of ignition for alcohol. In this system, specific fuel consumption depends on glow-plug positions and temperatures. It must be noted that the temperature of glow plugs must vary with load. However, the glow plug becomes inefficient at a high load. In addition, the specific fuel consumption is higher than that of diesel.

8.2.7 *Torque Tests*

Peak torque applies less to biodiesel fuels than it does to D2 fuel but occurs at lower engine speed and generally its torque curves are flatter. Testing includes the power and torque of the methyl esters and diesel fuel and ethyl esters *vs.* diesel fuel. Biodiesels on average decrease power by 5% compared to diesel fuel at a rated load (Demirbas, 2005).

8.2.8 *Spark Ignition*

When a spark plug is used, a diesel engine can be converted into an Otto cycle engine. In this case, the compression ratio should be reduced from 16:1 to 10.5:1. There are two types of this kind of conversion. (a) In the first type, the original fuel injection system is maintained. Alcohol requires additives for lubricity. In addition, both distributor and spark plug need to be installed, thus leading to a high cost of conversion. It is critical to adjust an ideal injection and spark time for this kind of conversion. (b) In the second type of conversion, the original fuel injection is eliminated. But a carburetor, spark plug, and distributor need to be installed, which increases the cost of conversion. In this conversion, spark timing is critical.

8.2.9 Oxidation

The effects of oxidative degradation caused by contact with ambient air (autoxidation) during long-term storage present a legitimate concern in terms of maintaining the fuel quality of biodiesel. Oxidative degradation reactions of biodiesel fuels were conducted in the laboratory under varying time and temperature conditions (Dunn, 2002). Results showed that reaction time significantly affects kinematic viscosity (v). With respect to increasing reaction temperature, v, acid value (AV), peroxide value (PV), and specific gravity (SG) increased significantly, whereas cold flow properties were minimally affected for temperatures up to 423 K. Antioxidants tert-butylhydroquinone (TBHQ) and α–tocopherol showed beneficial effects on retarding the oxidative degradation of methyl soyate (biodiesel) under the conditions of this study. Results indicated that v and AV have the best potential as parameters for timely and easy monitoring of biodiesel fuel quality during storage (Dunn, 2002). TBHQ is the most effective antioxidant for highly unsaturated vegetable oils and many animal fats.

The chemical compositions of fatty acids of sunflower seed oil and biodiesel are presented in Table 8.2. The content of linoleic acid in sunflower seed oil is 72.4%, while linoleic acid accounts for 62.5% of the total fatty acids in biodiesel. The proportion of linoleic acid is lower in biodiesels obtained by the supercritical methanol transesterification method. Partial degradation of the linoleic acid may occur in the supercritical methanol transesterification method due to high temperature (<513 K). In study the high temperature had a much greater effect on the linoleic acid, as a polyunsaturated fatty acid, than it did on saturated and monosaturated fatty acids.

Table 8.2 Chemical compositions of fatty acids in sunflower seed oil and biodiesel by supercritical methanol method

Fatty acid, as methyl ester	Vegetable oil	Biodiesel
Myristic (14:0)	0.1	0.1
Palmitic (16:0)	6.7	8.1
Stearic (18:0)	2.9	3.7
Oleic (9c–18:1)	17.7	25.4
Linoleic (9c, 12c–18:2)	72.4	62.5
Others	0.2	0.2

References

Demirbas, A. 2003. Biodiesel fuels from vegetable oils via catalytic and non-catalytic supercritical alcohol transesteri.cations and other methods: a survey. Energy Convers Mgmt 44:2093–2109.

Demirbas, A. 2005. Biodiesel production from vegetable oils via catalytic and non-catalytic supercritical methanol transesterification methods. Progress in Energy Combus Sci 31: 466–487.

Demirbas, A. 2006a. Theoretical heating values and impacts of pure compounds and fuels. Energy Sour Part A 28:459–467.

Demirbas A. 2006b. Biodiesel production via non-catalytic SCF method and biodiesel fuel characteristics. Energy Convers. Mgmt 47:2271–2282.

Dunn, R.O. 2002. Effect of oxidation under accelerated conditions on fuel properties of methyl soyate (biodiesel). JAOCS 79:915–920.

Goering, C.E., Camppion, R.N., Schwab, A.W., Pryde, E.H. 1982. In: Vegetable oil fuels, proceedings of the international conference on plant and vegetable oils as fuels, Fargo, North Dakota. American Society of Agricultural Engineers, St. Joseph, MI 4:279–286.

Isigigur, A., Karaosmonoglu, F., Aksoy, H.A. 1994. Methyl ester from safflower seed oil of Turkish origin as a biofuel for diesel engines. Appl Biochem Biotechnol 45/46:103–112.

Jenkins, B.M., Baxter, L.L., Miles, Jr., T.R., Miles, T.R. 1998. Combustion properties of biomass. Fuel Process Technol 54:17–46.

Kusdiana, D., Saka, S. 2001. Kinetics of transesterification in rapeseed oil to biodiesel fuels as treated in supercritical methanol. Fuel 80:693–698.

Ma, F., Hanna, M.A. 1999. Biodiesel production: a review. Bioresour Technol 70:1–15.

Peterson, C.L., Korus, R.A., Mora, P.G., Madsen, J.P. 1987. Fumigation with propane and transesterification effects on injector coking with vegetable oils. Trans ASAE 30:28–35.

Schwab, A.W., Bagby, M.O., Freedman, B. 1987. Preparation and properties of diesel fuels from vegetable oils. Fuel 66:1372–1378.

Shay, E.G. 1993. Diesel fuel from vegetable oils: status and opportunities. Biomass Bioenergy 4:227–242.

Chapter 9
Global Renewable Energy and Biofuel Scenarios

Projections are important tools for long-term planning and policy settings. Renewable energy sources that use indigenous resources have the potential to provide energy services with zero or almost zero emissions of both air pollutants and greenhouse gases. Renewable energy is a promising alternative solution because it is clean and environmentally safe. Currently, renewable energy sources supply 14% of the total world energy demand. Approximately half of the global energy supply will come from renewables in 2040. Photovoltaic (PV) systems and wind energy will play an important role in the energy scenarios of the future. The most significant developments in renewable energy production are observed in PVs (from 0.2 to 784 Mtoe) and wind energy (from 4.7 to 688 Mtoe) between 2001 and 2040.

Biofuels are expected to reduce dependence on imported petroleum with its associated political and economic vulnerability, reduce greenhouse gas emissions and other pollutants, and revitalize the economy by increasing demand and prices for agricultural products. Although most attention focuses on ethanol, interest in biodiesel is also increasing. Rapeseed is the primary oil used to make European biodiesel. Currently, biodiesel use is particularly strong in Germany. Biodiesel is primarily produced from soybeans in the United States. The European Union has chosen biodiesel as its main renewable liquid fuel. Fuel use of ethanol in the European Union is much less important. Low European corn production and a high proportion of diesel engines compared to the United States make biodiesel a more attractive alternative in the European Union.

For fuels produced from biomass, various conversion routes are available that follow from the different types of biomass feedstocks. These routes include direct conversion processes such as extraction of vegetable oils followed by esterification (biodiesel), fermentation of sugar-rich crops (ethanol), pyrolysis of wood (pyrolysis oil derived diesel equivalent), and hydrothermal upgrading (HTU) of wet biomass (HTU-oil-derived diesel equivalent). Another possibility is to produce liquid biofuels (methanol, DME, Fischer-Tropsch liquids) from synthesis gas, which results from the gasification of biomass. At present, the biofuel-producing countries in the European Union only have a small share in the global production of biofuels, namely, a little less than 6%. Most of the global biofuel production consists of ethanol. The main ethanol producers are the USA and Brazil. However, Europe is the most important producer of biodiesel on the global market.

There are several reasons why biodiesel is a more advantegious liquid fuel than of ethyl alcohol. First, production technology is rapidly evolving. Ethanol production has expanded rapidly through refinement of the enzyme process. Supporters of biodiesel expect similar advancements as their production equipment becomes more sophisticated and refined. Second, biodiesel has considerable market potential.

Biomass is currently the most used renewable energy source and will continue to be so in the foreseeable future. The potential of sustainable large hydro energy is quite limited to certain regions in the world. The potential for small hydro (<10 MW) power is still significant and take on increased significance in the future. PV systems and wind energy technologies have seen annual growth rates of more than 30% in recent years that will become more significant in the future. Geothermal and solar thermal sources will be more important energy sources in the future. PV will then be the largest renewable electricity source with a production of 25.1% of global power generation in 2040.

It has been estimated that known petroleum reserves will be depleted in less than 50 years at the present rate of consumption (Sheehan *et al.*, 1998). Oil is a finite resource. Various studies put the date of the global peak in oil production between 1996 and 2035. In developed countries there is a growing trend toward employing modern technologies and efficient bioenergy conversion using a range of biofuels, which are becoming cost competitive with fossil fuels (Puhan *et al.*, 2005).

Energy production from fossil fuels leads to high greenhouse gas emissions. Global use of fossil fuels is rapidly resulting in critical environmental problems throughout the world. The availability of fossil fuels, renewable energy sources, and nuclear sources strongly affects current and future energy supply systems. Declining availability of fossil fuels may cause increases in fuel prices and declining security of energy supplies in the current energy system. Limited availability of fossil fuels and renewable energy sources on a local scale may affect the sustainability of future energy systems. If the global growth rate of about 2% a year of primary energy use continues, it will mean a doubling of energy consumption by 2035 relative to 1998, and a tripling by 2055 (UNDP, 2000).

Global and regional availability has been estimated for wind energy, solar energy, especially PV cells, and biomass, together with a description of factors influencing this availability. In the long term, countries with surplus biomass potential could become exporters of bioenergy.

Renewable energy sources, or renewables, contributed 2% of the world's energy consumption in 1998, including 7 exajoules from modern biomass and 2 exajoules for all other renewables (UNDP, 2000). Renewables are clean or inexhaustible and primary energy resources. However, renewable technologies like water and wind power probably would not have provided the same fast increase in industrial productivity as fossil fuels did (Edinger and Kaul, 2000).

Oil and gas are expected to continue to be important sources of energy (Nakicenovic *et al.*, 1998). The share of renewable energy sources is expected to increase very significantly (to 30 to 80% in 2100). Hydropower and traditional biomass are already important factors in the world's energy mix, contributing about 18% of the total world energy requirements, whereas renewables contribute only about 2% of the present world primary energy use. Biomass, wind, and geothermal energy are commercially competitive and are making relatively fast progress (Fridleifsson, 2001).

9.1 Global Renewable Energy Sources

Renewable energy sources that use indigenous resources have the potential to provide energy services with zero or almost zero emissions of both air pollutants and greenhouse gases. Currently, renewable energy sources supply 14% of the total world energy demand. Large-scale hydropower supplies 20% of global electricity. Renewable resources are more evenly distributed than fossil and nuclear resources (Demirbas, 2006a). Renewable energy scenarios depend on environmental protection, which is an essential characteristic of sustainable developments.

For biomass resources, several potential sources may be used. Biomass resources are agricultural and forest residues, algae and grasses, animal manure, organic wastes, and biomaterials (Hoogwijk, 2004; Demirbas, 2007a). The supply is dominated by traditional biomass used for cooking and heating, especially in rural areas of developing countries. World production of biomass, mostly wild plant growth, is estimated at 146 billion metric tons a year (Cuff and Young, 1980). Worldwide biomass ranks fourth as an energy resource, providing *ca.* 14% of the world's energy needs (Hall *et al.*, 1992; Ramage and Scurlock, 2001).

Biomass now represents only 3% of primary energy consumption in industrialized countries. However, much of the rural population in developing countries, which represents about 50% of the world's population, is reliant on biomass, mainly in the form of wood, for fuel (Ramage and Scurlock, 2001). Large-scale hydropower provides about one quarter of the world's total electricity supply, virtually all of Norway's electricity, and more than 40% of the electricity used in developing countries. The technically usable world potential of large-scale hydro is estimated to be over 2200 GW.

There are two small-scale hydropower systems: microhydropower systems (MHP), with capacities below 100 kW, and small hydropower systems (SHP), with capacity between 101 kW and 1 MW. In the developing countries, considerable potential still exists, but large hydropower projects may face financial, environmental, and social constraints (UNDP, 2000).

Geothermal energy for electricity generation has been produced commercially since 1913, and for four decades on the scale of hundreds of megawatts both for electricity generation and direct use. The utilization has increased rapidly during the last three decades. In 2000, geothermal resources have been identified in over

80 countries, and there are quantified records of geothermal utilization in 58 countries in the world.

Geothermal energy is clean, cheap, and renewable and can be utilized in various forms such as space heating and domestic hot water supply, CO_2 and dry-ice production processes, heat pumps, greenhouse heating, swimming and balneology (therapeutic baths), industrial processes, and electricity generation. The main types of direct use are bathing, swimming and balneology (42%), space heating (35%), greenhouses (9%), fish farming (6%), and industry (6%) (Fridleifsson, 2001).

One of the most abundant energy sources on the surface of the earth is sunlight. Today, solar energy makes a tiny contribution to world total primary energy supply, less than 1% (Ramachandra, 2007).The potential of solar energy—passive solar heat, collectors for hot water for example, and PV power—is tremendous.

Following the oil crises of the 1970s, energy experts began to explore whether solar-based power generation held potential as an alternative to petroleum-based fuels. Development of solar power has progressed considerably since then, yet its record of performance has been mixed, and it has not come into widespread use in either industrialized or developing countries.

PV systems, other than solar home heating systems, are used for communication, water pumping for drinking and irrigation, and electricity generation. The total installed capacity of such systems is estimated at about 1000 kW. A solar home heating system is a solar PV system with a maximum capacity of 40 W. These systems are installed and managed by a household or a small community (Garg and Datta, 1998).

Like wind power markets, PV markets have seen rapid growth and costs have fallen dramatically. The total installed capacity of such systems is estimated at about 1000 kW. Solar PV is growing fast; the PV and grid-connected wind installed capacities are growing at a rate of 30% a year (Demirbas, 2005). Wind energy is a significant resource; it is safe, clean, and abundant. Wind energy is an indigenous supply permanently available in virtually every nation in the world. Using the wind to produce electricity by turning blades on a wind turbine is known as wind energy or wind power. More recently large wind turbines have been designed that are used to generate electricity. Wind as a source of energy is non-polluting and freely available in many areas. As wind turbines are becoming more efficient, the cost of the electricity they generate is falling.

Wind power in coastal and other windy regions is promising as well. Today there are wind farms around the world. Production of wind-generated electricity has risen from practically zero in the early 1980s to more than 7.5 TWh per year in 1995. Cumulative generating capacity worldwide topped 6500 MW in late 1997 (Demirbas, 2005). Wind energy is abundant, renewable, widely distributed, and clean and mitigates the greenhouse effect if it is used to replace fossil-fuel-derived electricity. Wind energy has limitations based on geography and meteorology, plus there may be political or environmental problems (*e.g.*, dead birds) with installing turbines (Garg and Datta, 1998). On the other hand, wind can cause air pollution by degradation and distribution of pieces of pollutants such as waste paper, straw, *etc*.

9.2 Renewable Energy Scenarios

World developments in the field of energy supply, after the oil crises of the 1970s and the oil crisis of 2004, are showing the way to more serious steps toward sustainability in strategic energy planning, the improvement of energy efficiency, and the rational use of energy. Renewable energy sources are increasingly becoming a key factor in this line of thought.

Renewable energy is a promising alternative solution because it is clean and environmentally safe. Renewable energy sources also produce lower or negligible levels of greenhouse gases and other pollutants when compared with the fossil energy sources they replace. Approximately half of the global energy supply will come from renewables in 2040 according to European Renewable Energy Council (EREC) (2006). The most significant developments in renewable energy production are observed in PVs (from 0.2 to 784 Mtoe) and wind energy (from 4.7 to 688 Mtoe) between 2001 and 2040.

Using fossil fuels as the primary energy source has led to a serious energy crisis and environmental pollution on a global scale. The limitations of solar power are site specific, intermittent, and, thus, not reliable for immediate supply. Using batteries to store an energy surplus for later consumption can resolve the time mismatch between energy supply and demand. The shortcomings of battery storage are low storage capacity, short equipment life, and generation of considerable solid and chemical waste. A system consisting of PV panels coupled with electrolyzers is a promising design to produce hydrogen (Ni *et al.*, 2006).

A detailed analysis of the technical, economic, and regulatory issues of wind power appears in the European Wind Energy Association (EWEA) Report: "Large scale integration of wind energy in the European power supply: analysis, issues and recommendations" published in December 2005. In 2005, worldwide capacity of wind-powered generators was 58,982 MW; although it currently produces less than 1% of worldwide electricity use, it accounts for 23% of electricity use in Denmark, 4.3% in Germany, and *ca.* 8% in Spain. Globally, wind power generation more than quadrupled between 1999 and 2005 according to the EWEA (2005).

Figure 9.1 shows the growth scenarios for global installed wind power (IEA, 2006). In 2004, the International Energy Agency (IEA) reference scenario projections for wind energy were updated to 66 GW in 2010, 131 GW in 2020, and 170 GW in 2030. The IEA reference scenario projections for wind energy were updated to 75 GW in 2010, 145 GW in 2020, and 202 GW in 2030. The IEA advanced strong growth scenario projected a wind energy market of 82 GW in 2010, 165 GW in 2020, and 250 GW in 2030.

The term biofuel refers to liquid or gaseous fuels for the transport sector that are predominantly produced from biomass. Biofuels mainly are bioethanol, biomethanol, biodiesel, biohydrogen, and biogas.

Biofuels are important because they replace petroleum-based fuels. Biofuels are generally considered as offering many advantages including sustainability,

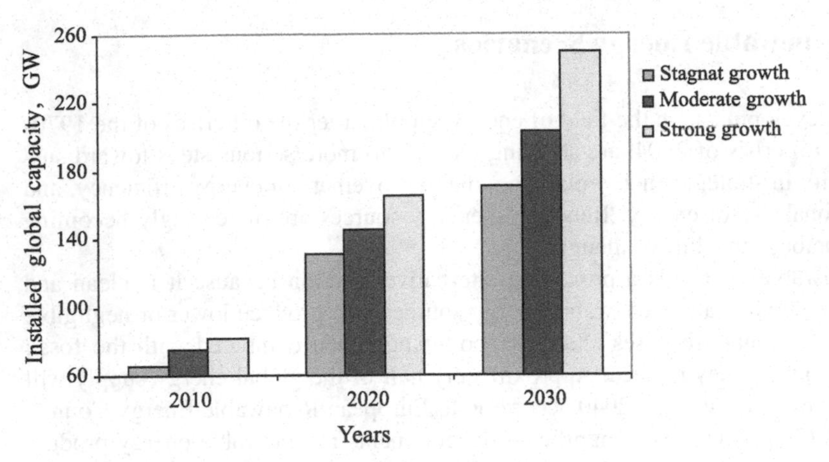

Fig. 9.1 Scenarios for global installed wind power

reduction of greenhouse gas emissions, regional development, new employment possibilities, and security of supply (Reijnders, 2006).

Biomass provides a number of local environmental benefits. Energy forestry crops have a much greater diversity of wildlife and flora than alternative land use, which is arable or pasture land. In industrialized countries, the main biomass processes used in the future are expected to be direct combustion of residues and wastes for electricity generation, bioethanol and biodiesel as liquid fuels, and combined heat and power production from energy crops. The future of biomass electricity generation lies in biomass integrated gasification/gas turbine technology, which offers high energy conversion efficiencies. Biomass will compete favorably with fossil mass for niches in the chemical feedstock industry. Biomass is a renewable, flexible, and adaptable resource. Crops can be grown to satisfy changing end use needs (Demirbas, 2006b).

In the future, biomass has the potential to provide a cost-effective and sustainable supply of energy while at the same time aiding countries in meeting their greenhouse gas reduction targets. By the year 2050, it is estimated that 90% of the world population will live in developing countries.

The importance of biomass in different world regions is given in Table 9.1. As shown in the table, the importance of biomass varies significantly across regions. In Europe, North America, and the Middle East, the share of biomass averages 2 to 3% of total final energy consumption, whereas in Africa, Asia, and Latin America, which together account for three quarters of the world's population, biomass provides a substantial share of the energy needs: a third on average, but as much as 80 to 90% in some of the poorest countries of Africa and Asia (*e.g.*, Angola, Ethiopia, Mozambique, Tanzania, Democratic Republic of Congo, Nepal, and Myanmar). Indeed, for large portions of the rural populations of developing countries, and for the poorest sections of urban populations, biomass is often the only available and affordable source of energy for basic needs such as cooking and heating (Demirbas, 2007b).

Table 9.1 The importance of biomass in different world regions

Region	Share of biomass in final energy consumption
Africa	62.0
Burundi	93.8
Ethiopia	85.6
Kenya	69.6
Somalia	86.5
Sudan	83.7
Uganda	94.6
South Asia	56.3
East Asia	25.1
China	23.5
Latin America	18.2
Europe	3.5
North America	2.7
Middle East	0.3

Table 9.2 Combustion and thermal efficiencies of certain biomass fuels

Biomass fuel	Combustion efficiency (%)	Overall thermal efficiency (%)
Wood	88.5–92.5	21–25
Wood char	83–87	13–15
Dung	85–89	10–12
Biogas	98.5–99.5	56–59
Agricultural residues	84–88	14–16

With relatively cold winters, biomass- or wood-fired domestic stoves are commonly used for space heating and cooking in rural areas. In our homes, we burn wood in stoves and fireplaces to cook meals and warm our residences. A number of portable biomass stoves have been designed, built, and tested over the last decade. The aim of the design is to burn the solid biomass fuels with a facility to have some control over power, maximize the efficiency of heat transfer into the vessel, minimize emissions, and reduce costs. The energy conversion efficiency of an efficient fuelwood stove is estimated to be 20 to 30% compared to about 10% efficiency of traditional fuelwood stoves. Therefore, efficient stoves can reduce fuelwood consumption significantly. Much of the research and development work carried out on biomass technologies for rural areas of developing countries has been based on the improvement of traditional stoves (Demirbas, 2007b). Combustion and thermal efficiencies of some biomass fuels used in traditional stoves are given in Table 9.2.

The combustion of biomass produces significantly less NO_x and SO_2 emissions than the burning of fossil fuels. Globally significant environmental benefits may result from using biomass for energy rather than fossil fuels. The greatest benefit is derived from substituting biomass energy for coal. The degree of benefit depends greatly on the efficiency with which the biomass is converted into electricity. There will be a considerable reduction in net carbon dioxide emissions that contribute to the greenhouse effect. For example, one dry ton of wood will displace 15 GJ of coal. The 15 GJ of coal will have the equivalent of 0.37 tons of carbon assuming the wood is converted at an efficiency of 25% (Tewfik, 2004).

According to the IEA7, scenarios developed for the USA and the EU indicate that near-term targets of up to 6% displacement of petroleum fuels with biofuels appear feasible using conventional biofuels, given available cropland. A 5% displacement of gasoline in the EU requires about 5% of available cropland to produce ethanol, while in the USA 8% is required. A 5% displacement of diesel requires 13% of US cropland, 15% in the EU (IEA, 2006).

The recent commitment by the US government to increase bioenergy threefold in 10 years has added impetus to the search for viable biofuels. The advantages of biofuels are the following: (a) they are easily available from common biomass sources, (b) they represent a carbon dioxide cycle in combustion, (c) they are environmentally friendly, and (d) they are biodegradable and contribute to sustainability (IEA, 2004).

The dwindling fossil fuel sources and the increasing dependency of the USA on imported crude oil have led to a major interest in expanding the use of bioenergy. In addition to the aforementioned recent commitment by the US government to increase bioenergy use, the EU has also adopted a proposal for a directive on the promotion of the use of biofuels with measures ensuring that biofuels account for at least 2% of the market for gasoline and diesel sold as transport fuel in 2005, increasing in stages to a minimum of 5.75% by the end of 2010 (Puppan, 2002). Bioethanol is a fuel derived from renewable sources of feedstock, typically plants such as wheat, sugar beet, corn, straw, and wood. Bioethanol is a petrol additive/substitute. Biodiesel is better than diesel fuel in terms of sulfur content, flash point, aromatic content, and biodegradability (Hansen *et al.*, 2005).

Solar PV systems hold great promise as a future energy source. PV will then be the largest renewable electricity source with a production of 25.1% of global power generation in 2040 (EWEA, 2005). PV systems, or photovoltaics, offer consumers the ability to generate electricity in a clean, quiet, and reliable way. These systems consist of a PV array, control and safety equipment, a battery bank, and, usually, an inverter. Because the source of light is the sun, they are often called solar cells. Therefore, the PV process produces electricity directly from sunlight. PV cells are made of semiconductor material. When light enters the cell, some of the photons from the light are absorbed by the semiconductor atoms, freeing electrons to flow through an external circuit and back into the cell. This flow of electrons is an electric current. Solar PV generators are networks of various interconnections between solar cells, diodes, cables, and other components. The efficiency of semiconductor electrodes exposed to concentrated sunlight has been studied, and it has been found that the batteries play a vital role in solar PV refrigeration systems. Despite breakthroughs in operational characteristics of various components of such systems, lead acid batteries continue to be the only viable electrical energy storage device as of date (Kattakayam and Srinivasan, 2004).

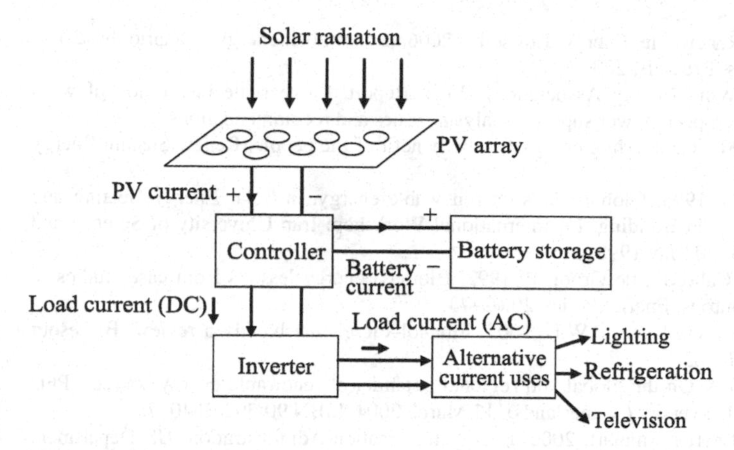

Fig. 9.2 Configuration of PV system with AC appliances

Solar PV energy as a renewable energy source has the greatest potential worldwide. However the development of this source of energy depends on ambitious political support.

Solar PV and diesel systems should be compared on the basis of life-cycle costs of providing the final services that the customer desires for a number of years (*e.g.*, household lighting, refrigeration, video). In particular, solar PV and diesel systems should not be compared on the basis of cost per kilowatt hour of electricity produced under the two systems because such a comparison fails to account for the major operational differences between solar PV and diesel systems. A comparison of solar PV and diesel systems on the basis of life-cycle costs shows that solar PV systems are marginally cheaper than diesel systems for households in remote rural areas. Figure 9.2 shows the configuration of PV systems with AC appliances (*e.g.*, household lighting, refrigeration, television, video).

References

Cuff, D.J., Young, W.J. 1980. US energy atlas. New York: Free Press/McMillan.

Demirbas, A. 2005. Potential applications of renewable energy sources, biomass combustion problems in boiler power systems and combustion related environmental issues. Prog Energy Combust Sci 31:171–192.

Demirbas, A. 2006a. Energy priorities and new energy strategies. Energy Edu Sci Technol 16: 53–109.

Demirbas, A. 2006b. Global biofuel strategies. Energy Edu Sci Technol 17:32–63.

Demirbas, A. 2007a. Production of biofuels from macroalgae and microalgae. Energy Edu Sci Technol 18:59–65.

Demirbas, A. 2007b. Combustion systems for biomass fuels. Energy Sources Part A 29:303–312.

Edinger, R., Kaul, S. 2000. Humankind's detour toward sustainability: past, present, and future of renewable energies and electric power generation. Renew Sustain Energy Rev 4:295–313.

EREC (European Renewable Energy Council). 2006. Renewable energy scenario by 2040, EREC Statistics, Brussels, 2006.

EWEA (European Wind Energy Association). 2005. Report: Large scale integration of wind energy in the European power supply: Analysis, issues and recommendations.

Fridleifsson, I.B. 2001. Geothermal energy for the benefit of the people. Renew Sustain Energy Rev 5:299–312.

Garg, H.P., Datta, G. 1998. Global status on renewable energy, in Solar Energy Heating and Cooling Methods in Building. In: International Workshop: Iran University of Science and Technology. 19–20 May 1998.

Hall, D.O., Rosillo-Calle, F., de Groot, P. 1992. Biomass energy lessons from case studies in developing countries. Energy Policy 20:62–73.

Hansen, A.C., Zhang, Q., Lyne, P.W.L. 2005. Ethanol–diesel fuel blends–a review. Bioresour Technol 96:277–285.

Hoogwijk, M.M. 2004. On the global and regional potential of renewable energy sources. PhD thesis, Utrecht University (Netherlands), 12 March 2004, ISBN 90-393-3640, 7.

IEA (International Energy Annual). 2000. Energy Information Administration, US Department of Energy, Washington, DC.

IEA (International Energy Agency). 2004. Biofuels for transport: An international perspective. 9, rue de la Fédération, 75739 Paris, cedex 15, France. www.iea.org.

IEA (International Energy Agency). 2006. Reference scenario projections. 75739 Paris, cedex 15, France.

Kattakayam, T.A, Srinivasan, K. 2004. Lead acid batteries in solar refrigeration systems. Renew Energy 29:1243–1250.

Nakicenovic, N., Grübler, A., McDonald, A. (eds.) 1998. Global Energy Perspectives. Cambridge University Press, Cambridge, UK.

Ni, M., Leung, M.K.H, Sumathy, K., Leung, D.Y.C. 2006. Potential of renewable hydrogen production for energy supply in HongKong. Int J Hydrogen Energy 31:1401–1412.

Puhan, S., Vedaraman, N., Rambrahaman, B.V., Nagarajan, G. 2005. Mahua (Madhuca indica) seed oil: a source of renewable energy in India. J Sci Ind Res 64:890–896.

Puppan, D. 2002. Environmental evaluation of biofuels. 2002. Periodica Polytechnica Ser Soc Man Sci 10:95–116.

Ramachandra, T.V. 2007. Solar energy potential assessment using GIS. Energy Edu Sci Technol 18:101–114.

Ramage, J., Scurlock, J. 1996. Biomass. In: Boyle, G. (ed.) Renewable Energy-Power for a Sustainable Future. Oxford University Press, Oxford.

Reijnders, L. 2006. Conditions for the sustainability of biomass based fuel use. Energy Policy 34:863–876.

Schobert, H.H., Song, C. 2002. Chemicals and materials from coal in the 21st century. Fuel 81:15–32.

Sheehan, J., Cambreco, V., Duffield, J., Garboski, M., Shapouri, H. 1998. An overview of biodiesel and petroleum diesel life cycles. A report by the US Department of Agriculture and Energy, Washington, DC, pp. 1–35.

Tewfik, S.R. 2004. Biomass utilization facilities and biomass processing technologies. Energy Edu Sci Technol 14:1–19.

UNDP (United Nations Development Programme). 2000. World Energy Assessment. Energy and the Challenge of Sustainability.

Chapter 10
The Biodiesel Economy and Biodiesel Policy

10.1 Introduction to the Biodiesel Economy

A number of technical and economic advantages of biodiesel fuel are that (1) it prolongs engine life and reduces the need for maintenance (biodiesel has better lubricating qualities than fossil diesel), (2) it is safer to handle, being less toxic, more biodegradable, and having a higher flash point, and (3) it reduces some exhaust emissions (although it may, in some circumstances, raise others) (Wardle, 2003).

Biodiesel is an efficient, clean, 100% natural energy alternative to petroleum-based fuels. Among the many advantages of biodiesel fuel are that it is safe for use in all conventional diesel engines, offers the same performance and engine durability as petroleum diesel fuel, is non-flammable and non-toxic, and reduces tailpipe emissions, visible smoke, and noxious fumes and odors (Chand, 2002). Biodiesel is better than diesel fuel in terms of sulfur content, flash point, aromatic content, and biodegradability (Martini and Schell, 1997). Figure 10.1 shows the biodiesel production of the European Union (1993–2005).

Commercial production of biodiesel began in the 1990s. Diesel mainly derived from rapeseed oil is the most common biodiesel available in Europe. The most common sources of oil for biodiesel production in the United States are soybean oil and yellow grease (primarily recycled cooking oil from restaurants). Among liquid biofuels, biodiesel derived from vegetable oils is gaining acceptance and market share as diesel fuel in Europe and the United States. By several important measures biodiesel blends perform better than petroleum diesel, but its relatively high production costs and the limited availability of some of the raw materials used in its production continue to limit its commercial application. Limiting factors of the biodiesel industry are feedstock prices, biodiesel production costs, crude oil prices, and taxation of energy products.

The economic benefits of a biodiesel industry would include value added to the feedstock, an increased number of rural manufacturing jobs, increased income taxes, increased investments in plant and equipment, an expanded manufacturing sector, an increased tax base from plant operations and income taxes, improvement

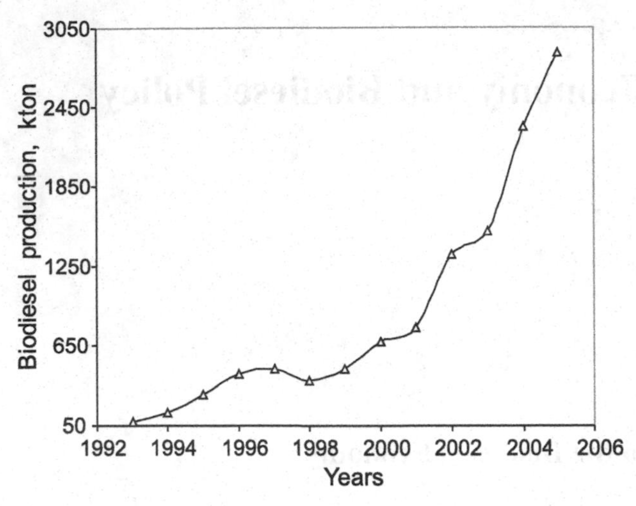

Fig. 10.1 Biodiesel production of the European Union (1993–2005)

in the current account balance, and reductions in health care costs due to improved air quality and greenhouse gas mitigation. The production and utilization of biodiesel is facilitated firstly through the agricultural policy of subsidizing the cultivation of non-food crops. Secondly, biodiesel is exempt from the oil tax. The European Union accounted for nearly 89% of all biodiesel production worldwide in 2005. By 2010, the United States is expected to become the world's largest single biodiesel market, accounting for roughly 18% of world biodiesel consumption, followed by Germany.

The economic advantages of biodiesel are that it reduces greenhouse gas emissions, helps to reduce a country's reliance on crude oil imports, and supports agriculture by providing new labor and market opportunities for domestic crops. In addition it enhances lubrication and is widely accepted by vehicle manufacturers (Palz *et al.*, 2002; Clarke *et al.*, 2003).

The major economic factor to consider with respect to the input costs of biodiesel production is the feedstock, which is about 80% of the total operating cost. Other important costs are labor, methanol, and catalyst, which must be added to the feedstock. In some countries, filling stations sell biodiesel more cheaply than conventional diesel.

The cost of biodiesel fuels varies depending on the base stock, geographic area, variability in crop production from season to season, the price of crude petroleum, and other factors. Biodiesel costs more than twice petroleum diesel. The high price of biodiesel is in large part due to the high price of the feedstock. However, biodiesel can be made from other feedstocks, including beef tallow, pork lard, and yellow grease.

Biodiesel is attracting increased attention from companies interested in commercial-scale production as well as the more usual home brew biodiesel user and the user of straight vegetable oil or waste vegetable oil in diesel engines. Biodiesel is commercially available in most oilseed-producing countries. Biodiesel is a technologically feasible alternative to petrodiesel, but nowadays biodiesel costs 1.5 to 3 times more than fossil diesel in developed countries. Biodiesel is more expensive than petrodiesel, though it is still commonly produced in relatively small quantities (in comparison to petroleum products and ethanol). The competitiveness of biodiesel to petrodiesel depends on the fuel taxation rates and policies. Generally, the production costs of biodiesel remain much higher than those of petrodiesel. Therefore, biodiesel is not competitive with petrodiesel under current economic conditions. The competitiveness of biodiesel relies on the price of the biomass feedstock and costs associated with the conversion technology.

The recent increase in the potential use of biodiesel is due not only to the number of plants but also to the size of the facilities used in its production. The tremendous growth in the biodiesel industry is expected to have a significant impact on the price of biodiesel feedstocks. This growth in the biodiesel industry will increase competition. An earlier evaluation of the potential feedstocks for biodiesel by Hanna *et al.* (2005) also identified the expected price pressures on biodiesel feedstocks. Fiscal incentives for biodiesel such as reductions in feedstock and processing costs and tax exemptions will be the key tool for enhancing the use of biodiesel as an alternative fuel for transport in the near future. Biodiesel possesses a number of promising characteristics, including reduction of exhaust emissions (Dunn, 2001). The advantages offered by biodiesel must be considered at levels beyond the agricultural, transport, and energy sectors only.

10.2 Economic Benefits of Biodiesel

Biodiesel is a renewable fuel manufactured from vegetable oils, animal fats, and recycled cooking oils. Biodiesel offers many benefits (USDE, 2006):

1. It is renewable.
2. It is energy efficient.
3. It displaces petroleum-derived diesel fuel.
4. It can be used in most diesel equipment with no or only minor modifications.
5. It can reduce global warming gas emissions.
6. It can reduce tailpipe emissions, including air toxins.
7. It is non-toxic, biodegradable, and suitable for sensitive environments.
8. It is made from either agricultural or recycled resources.

In recent years, the importance of non-food crops has increased significantly. The need to grow non-food crops under a compulsory set-aside scheme creates an opportunity to increase biodiesel production, but this is not an appropriate instrument to promote non-food production.

An important factor that is not usually considered when calculating the costs and benefits of industrial feedstock materials is the macroeconomic effect associated with domestically produced renewable energy sources.

The industrial production of biodiesel in Europe is not profitable without government subsidies. In the early 1990s, when the biodiesel industry was incurring heavy startup costs, the contract price for rapeseed used in biodiesel manufacture was much lower than for rapeseed used for food. This was possible because rapeseed for biodiesel was grown on land where the growing of crops for food and feed were not allowed. Farmers, with no better option, were willing to produce oilseeds (primarily rapeseed) under contract for less than the food use market price. For a farmer to raise oilseeds on fallow land, they must sign a contract with a buyer who registers the contract with the appropriate national government agency. However, since 1998, contract prices have been similar to cash prices.

Blends of up to 20% biodiesel mixed with petrodiesel fuels can be used in nearly all diesel equipment and are compatible with most storage and distribution equipment. Higher blends, even B100, can be used in many engines built with little or no modification. Transportation and storage, however, require special management. Material compatibility and warranty issues have not been resolved with higher blends.

Biodiesel has become more attractive recently because of its environmental benefits. The cost of biodiesel, however, is the main obstacle to commercialization of the product. With cooking oils used as raw material, the viability of a continuous transesterification process and the recovery of high-quality glycerol as a biodiesel byproduct are primary options to be considered to lower the cost of biodiesel (Ma and Hanna, 1999; Zhang *et al.*, 2003).

Most of the biodiesel that is currently made uses soybean oil, methanol, and an alkaline catalyst. Methanol is preferred because it is less expensive than ethanol (Graboski and McCormick, 1998). A base catalyst is preferred in transesterification because the reaction is quick and thorough. It also occurs at lower temperature and pressure than other processes, resulting in lower capital and operating costs for the biodiesel plant. The high value of soybean oil as a food product makes the production of a cost-effective fuel very challenging. However, there are large amounts of low-cost oils and fats, such as restaurant waste and animal fats, that could be converted into biodiesel. The problem with processing these low-cost oils and fats is that they often contain large amounts of free fatty acids that cannot be converted into biodiesel using an alkaline catalyst (Canakci and Van Gerpen, 2001).

The total energy use for biodiesel production with the common method is 17.9 MJ/l biodiesel. Transesterification alone consumes 4.3 MJ/l, while our calculations show that the supercritical methanol method requires as much as 3.3 MJ/l, or an energy reduction of 1.0 MJ for each liter of biodiesel fuel. In the common catalyzed method, mixing is significant during the reaction. In our method, since the reactants are already in a single phase, mixing is not necessary. Since our process is much simpler, particularly in the purification step, which only requires the removal of unreacted methanol, it is further expected that a cost reduction of *ca.* 20% can be realized from the transesterification process. Therefore, the production cost for biodiesel fuel from rapeseed oil turns out to be US$0.59/l, compare to US$0.63/l for the common catalyzed method (Saka and Kusdiana, 2001).

The brake power of biodiesel was nearly the same as with petrodiesel, while the specific fuel consumption was higher than that of petrodiesel. Carbon deposits inside the engine were normal, with the exception of intake valve deposits. The results showed that transesterification treatment decreased the injector coking to a level significantly lower than that observed with petrodiesel (Demirbas, 2003). Although most researchers agree that vegetable oil ester fuels are suitable for use in CIEs, a few contrary results have also been obtained. The results of these studies point out that most vegetable oil esters are suitable as diesel substitutes but that more long-term studies are necessary for commercial use to become practical.

10.3 Biodiesel Costs

The estimated costs for biodiesel can be split up into fixed and variable costs. Fixed costs come from extracting the vegetable oil from seed and processing this vegetable oil into biodiesel. These costs include manufacturing, capital, and labor costs. Glycerol and protein meal for livestock feed are byproducts that might help to offset the cost of biodiesel production. The sale of these byproducts is considered fixed income.

The low production costs of crude oil derivatives are another crucial handicap for biodiesel marketing. In this sense, the continuous increase in crude oil prices brings the costs of producing petroleum-based diesel more in line with those of petroleum-based diesel, converting this difference from a handicap into a potential opportunity for promoting biodiesel use.

The cost of biodiesel is higher than diesel fuel. Currently, there are seven producers of biodiesel in the United States. Pure biodiesel (100%) sells for about US$1.50 to US$2.00 per gallon before taxes. Fuel taxes add approximately US$0.50 per gallon.

Due to the underutilization of triglyceride processing equipment for rapeseed, soybean, sunflower oils, or animal fats are used as the only feedstock. Examining the feedstocks against each other does not provide for a valid comparison. Therefore, a comparison should be made based on how much of the processing equipment is actually utilized. If only 38% of the processing equipment is allocated to the costs of production when rapeseed or sunflower oils are used, the cost per gallon of biodiesel decreases. Also, the cost per gallon of biodiesel from animal fats decreases if none of the preprocessing capital costs are allocated to the total costs of production. The costs per gallon for the feedstocks are listed in Table 10.1.

The other reactant in producing biodiesel is methanol. Methanol is a readily available commodity in the chemical industry. It is produced from natural gas. Methanol is valued at around €250 to €280 per ton, but the price varies with the

Table 10.1 Costs per gallon for four feedstocks

Feedstock	Cost ($/gallon)
Animal fats 1.35	1.35
Rapeseed oil 1.46	1.46
Sunflower oil 2.35	2.35
Soybean oil 1.26	1.26

price of natural gas. A catalyst is necessary for the reaction, but the catalyst used varies from one biodiesel manufacturing plant to another. Sodium hydroxide, potassium hydroxide, and sulfuric acid are three commonly used catalysts.

The feedstock cost of the oil or grease is the largest single component of biodiesel production costs. Biodiesel from yellow grease is closer to being cost-competitive with petroleum diesel. Yellow grease is much less expensive than soybean oil, but its supply is limited.

Glycerine is produced as a coproduct with methyl ester, and it has economic value. One part of glycerine is produced for every ten parts of methyl ester. In the typical biodiesel plant, crude glycerine is produced, which is about 80% pure, water being the principal impurity. It has a price of about €500 per ton. With some additional investment, crude glycerine can be converted into pharmaceutical glycerine at 99.5% purity, which carries a price of €1030 per ton.

Current production costs of rapeseed methyl ester (RME) amount to *ca.* €0.50 per liter (or 15 €/GJ). These costs depend on the prices of the biomass used and the size and type of the production plant. The short-term investment costs for a 400-MWth plant are about 150 €/kWth. In the long term, these costs may decrease by about 30% for a larger sized plant with a thermal input capacity of 1000 MWth, assuming economies of scale. Other important factors determining the production costs of RME are the yield and value of the byproducts of the biodiesel production process, such as oilseed cake (a protein-rich animal feed) and glycerine (used in the production of soap and as a pharmaceutical medium). Longer-term projections indicate a future decrease in RME production costs by more than 50%, down to *ca.* €0.20 per liter (or around 6 €/GJ). However, to provide the amount of energy equivalent to 1 l of petroleum-derived diesel, a larger amount of RME is needed due to its lower energy content (Faaij and Hamelinck, 2001; CEC, 2001).

A review of 12 economic feasibility studies shows that the projected costs for biodiesel from oilseed or animal fats have a range of US$0.30 to 0.69/l, including meal and glycerine credits and assuming reduced capital investment costs by having the crushing and/or esterification facility added onto an existing grain or tallow facility. Rough projections of the cost of biodiesel from vegetable oil and waste grease are, respectively, US$0.54 to 0.62/l and US$0.34 to 0.42/l. With pretax diesel priced at US$0.18/l in the USA and US$0.20 to 0.24/l in some European countries, biodiesel is thus currently not economically feasible, and more research and technological development will be needed (Demirbas, 2003; Bender, 1999).

10.4 General Biodiesel Policy

In general, energy policy includes issues of energy production, distribution, and consumption. It is the manner in which a given entity has decided to address these issues. The attributes of energy policy may include international treaties, legislation on commercial energy activities (trading, transport, storage, *etc.*), incentives for investment, guidelines for energy production, conversion, and use (efficiency and emission standards), taxation and other public policy techniques, energy-related research and development, energy economy, general international trade agreements and marketing, energy diversity, and risk factors for possible energy crisis. Current energy policies also address environmental issues including environmentally friendly technologies to increase energy supplies and encourage cleaner, more efficient energy use, air pollution, the greenhouse effect (mainly reducing carbon dioxide emissions), global warming, and climate change (Demirbas, 2006a).

Many farmers who raise oilseeds use a biodiesel blend in tractors and equipment as a matter of policy to foster production of biodiesel and raise public awareness. It is sometimes easier to find biodiesel in rural areas than in cities. Additional factors must be taken into account such as the fuel equivalent of the energy required for processing, the yield of fuel from raw oil, the return on cultivating food, and the relative cost of biodiesel versus petrodiesel. Some nations and regions that have pondered transitioning fully to biofuels have found that doing so would require immense tracts of land if traditional crops were used.

While well intentioned, policymakers are on the wrong track for promoting biodiesel. Biodiesel policy should be based firmly on the philosophy of freedom. Given the history of petroleum politics, it is imperative that today's policy decisions ensure a free market for biodiesel. Producers of all sizes must be free to compete in this industry, without farm subsidies, regulations, and other interventions skewing the playing field. The production and utilization of biodiesel is facilitated firstly through the agricultural policy of subsidizing the cultivation of non-food crops. Secondly, biodiesel is exempt from the oil tax.

The Common Agricultural Policy Reform of 1992 established crop-specific payments per hectare to compensate for the reduction or abolition of institutional prices. The reference set-aside share is currently 10%, but the applied set-aside rates have been adapted year by year, taking into account market forecasts. Furthermore, farmers are flexible in the management of their set-aside obligations.

The source for biodiesel production is chosen according to the availability in each region or country. Any fatty acid source may be used to prepare biodiesel, but most scientific studies take soybean as a biodiesel source. Since the prices of edible vegetable oils, such as soybean oil, are higher than that of diesel fuel, waste vegetable oils and non-edible crude vegetable oils are preferred as potential low-priced biodiesel sources. Low-quality underused feedstocks have been used to produce biodiesel (Pinto *et al.*, 2005).

10.5 European Biofuel Policy

At present, biofuels are not competitive due to the relatively low price of crude oil. Biofuels are usually produced and used locally. In recent years, this pattern has been changed in northern Europe by industrial and large scale-use of different forms of biofuels. The trade situation has come about as a result of means of control on waste and energy (Demirbas, 2006b).

European research and testing indicate that, as a diesel-fuel substitute, biodiesel can replace petroleum diesel. Biodiesel, produced mainly from rapeseed or sunflower seed, comprises 80% of Europe's total biofuel production. The European Union accounted for nearly 89% of all biodiesel production worldwide in 2005. Germany produced 1.9 billion liters, or more than half the world total. Other countries with significant biodiesel markets in 2005 included France, the United States, Italy, and Brazil. All other countries combined accounted for only 11% of world biodiesel consumption in 2005. In Germany biodiesel is also sold at a lower price than fossil diesel fuel. Biodiesel is treated like any other vehicle fuel in the UK. The European Union has set the goal of obtaining 5.75% of transportation fuel needs from biofuels by 2010 in all member states. Many countries have adopted various policy initiatives. Specific legislation to promote and regulate the use of biodiesel is in force in Germany, Italy, France, Austria, and Sweden. New and large single markets for biodiesel are expected to emerge in China, India, and Brazil (EBB, 2004; Pinto *et al.*, 2005; Hanna *et al.*, 2005).

The general EU policy objectives considered most relevant to the design of energy policy are (1) the competitiveness of the EU economy, (2) the security of energy supplies, and (3) environmental protection. All renewable-energy policies should be evaluated on the basis of the contributions they make to these goals. Current EU policies on alternative motor fuels focus on the promotion of biofuels. In the European Commission's view mandating the use of biofuels will (a) improve energy supply security, (b) reduce greenhouse gas (GHG) emissions, and (c) boost rural incomes and employment. Elements of the European biofuels policy are (EC, 2003):

- A communication presenting the action plan for the promotion of biofuels and other alternative fuels in road transport.
- A directive on the promotion of biofuels for transport that requires an increasing proportion of all diesel and gasoline sold in the member states to be biofuel.
- Biofuels taxation, which is part of the large draft directive on the taxation of energy products and electricity, proposing to allow member states to apply differentiated tax rates in favor of biofuels.

The EU has also adopted a proposal for a directive on the promotion of the use of biofuels with measures ensuring that biofuels account for at least 2% of the market for gasoline and diesel sold as transport fuel by the end of 2005, increasing in stages to a minimum of 5.75% by the end of 2010 (Hansen *et al.*, 2005). The French Agency for Environment and Energy Management (ADEME) estimates that the 2010 objective would require industrial rapeseed plantings to increase from the current 3 million ha in the EU to 8 million ha (USDA, 2003).

In Germany, the current program of development of the biodiesel industry is not a special exemption from EU law but rather is based on a loophole in the law. The motor fuels tax in Germany is based on mineral fuel. Since biofuel is not a mineral fuel, it can be used for motor transport without being taxed. Unlike France and Italy, where biodiesel is blended with mineral diesel, biodiesel sold in Germany is pure, or 100%, methyl ester. There is no mineral tax on biodiesel in Germany, so when diesel prices were high and vegetable oil prices were low, biodiesel became very profitable. Additionally, there have been no restrictions on the quantity of biodiesel that can be exempted from the mineral fuel tax, so there has been a huge investment in biodiesel production capacity (USDA, 2003).

References

Bender, M. 1999. Economic feasibility review for community-scale farmer cooperatives for biodiesel. Bioresour Technol 70:81–87.

Chand, N. 2002. Plant oils—fuel of the future. J Sci Ind Res 61:7–16.

Canakci, M., Van Gerpen, J. 2001. Biodiesel production from oils and fats with high free fatty acids. Trans ASAE 44:1429–1436.

CEC (Commission of the European Communities). 2001. Communication from the Commission to the European Parliament, the Council, the Economic and Social Committee and the Committee of the Regions on alternative fuels for road transportation and on a set of measures to promote the use of biofuels. COM (2001) 547, 7 November 2001. Brussels.

Clarke, L.J., Crawshaw, E.H., Lilley, L.C. 2003. Fatty acid methyl esters (FAMEs) as diesel blend component. In: 9th Annual Fuels & Lubes Asia Conference and Exhibition, Singapore, 21–24 January 2003.

Demirbas, A. 2003. Biodiesel fuels from vegetable oils via catalytic and non-catalytic supercritical alcohol transesteri.cations and other methods: a survey. Energy Convers Mgmt 44:2093–2109.

Demirbas, A. 2006a. Energy priorities and new energy strategies. Energy Edu Sci Technol 16:53–109.

Demirbas, A. 2006b. Global biofuel strategies. Energy Edu Sci Technol 17:27–63.

Dunn, R.O. 2001. Alternative jet fuels from vegetable oils. Trans ASAE 44: 1155–1157.

EBB (European Biodiesel Board). 2004. EU: Biodiesel industry expanding use of oilseeds, Brussels.

EC (European Commission). 2003. Renewable energies: a European policy. Promoting Biofuels in Europe. European Commission, Directorate General for Energy and Transport, B-1049 Brussels, Belgium.

EPA (US Environmental Protection Agency). 2002. A comprehensive analysis of biodiesel impacts on exhaust emissions. Draft Technical Report, EPA420-P-02-001, October 2002.

Faaij, A., Hamelinck, C. 2001. Future prospects of methanol and hydrogen from biomass. University of Utrecht, Copernicus Institute, Department of Science, Technology and Society. The Netherlands.

Graboski, M.S., McCormick, R.L. 1998. Combustion of fat and vegetable oil derived fuels in diesel engines. Prog Energy Combust Sci 24:125–164.

Hanna, M.A., Isom, L., Campbell, J. 2005. Biodiesel: current perspectives and future. J Sci Ind Res 64:854-857.

Hansen, A.C., Zhang, Q., Lyne, P.W.L. 2005. Ethanol–diesel fuel blends—a review. Bioresour Technol 96:277–285.

Ma, F., Hanna, M.A. 1999. Biodiesel production: a review. Bioresour Technol 70:1–15.

Martini, N., Schell, S. 1997. Plant oils as fuels: present state of future developments. In: Plant Oils as Fuels – Present State of Science and Future Developments, Proceedings of the Symposium held in Potsdam, Germany, Springer, Berlin, p.6.

Palz, W., Spitzer, J., Maniatis, K., Kwant, N., Helm, P., Grassi, A. 2002. In: Proceeedings of the 12th International European Biomass Conference; ETA-Florence, WIP-Munich, Amsterdam, The Netherlands.

Pinto, A.C., Guarieiro, L.L.N., Rezende, M.J.C., Ribeiro, N.M., Torres, E.A., Lopes, W.A., Pereira, P.A.P., Andrade, J.B. 2005. Biodiesel: an overview. J Brazil Chem Soc 16: 1313-1330.

Saka, S., Kusdiana, D. 2001. Biodiesel fuel from rapeseed oil as prepared in supercritical methanol. Fuel 80:225–231.

Schumacher, S., Borgelt, L.G., Fosseen, S.C., Goetz, D., Hires, W.G. 1996. Heavy-duty engine exhaust emission test using methyl ester soybean oil/diesel fuel blends. Bioresour Technol 57:31–36.

Sensoz, S., Angin, D., Yorgun, S. 2000. Influence of particle size on the pyrolysis of rapeseed (*Brassica napus L.*): fuel properties of bio-oil. Biomass Bioenergy 19:271–279.

Sheehan, J., Cambreco, V., Duffield, J., Garboski, M., Shapouri, H. 1998. An overview of biodiesel and petroleum diesel life cycles. A report by the US Department of Agriculture and Energy, pp. 1–35.

Sheehan, J., Dunahay, T., Benemann, J., Roessler, P. 1998. A Look Back at the U.S. Department of Energy's Aquatic Species Program—Biodiesel from Algae. National Renewable Energy Laboratory (NREL) Report: NREL/TP-580-24190. Golden, CO.

USDE (US Department of Energy). 2006. Biodiesel handling and use guidelines. DOE/GO-102006-2358, 3rd edn., Oak Ridge, TN.

USDA (United States Department of Agriculture). 2003. Production Estimates and Crop Assessment Division of Foreign Agricultural Service. EU: Biodiesel Industry Expanding Use of Oilseeds. http://www.biodiesel.org/resources/reportsdatabase/reports/gen/20030920_gen330.pdf

Wardle, D.A. 2003. Global sale of green air travel supported using biodiesel. Renew Sustain Energy Rev 7:1–64.

Zhang, Y., Dub, M.A., McLean, D.D., Kates, M. 2003. Biodiesel production from waste cooking oil: 2. Economic assessment and sensitivity analysis. Bioresour Technol 90:229–240.

Index

H

M

O

P

R

S

T

V